Activities and Projects

for the

Freeman Statistics Series

Ron Millard
Shawnee Mission South High School
Graceland University

John C. Turner
U. S. Naval Academy

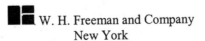 W. H. Freeman and Company
New York

ISBN-13: 978-0-7167-9145-4
ISBN-10: 0-7167-9145-5

Printed in the United States of America

Fifth Printing

W. H. Freeman and Company
41 Madison Avenue
New York, NY 10010
Houndmills, Basingstoke RG21 6XS, England

www.whfreeman.com

Table of Contents

Statistical Activities

Ron Millard
Shawnee Mission South High School
Graceland University

Preface to Statistical Activities

Students, statistics, and learning are the elements that I have combined in the classroom by using projects. All students do not learn in the same manner, and projects that focus on specific learning objectives provide a modality for some to discover statistical concepts that may not have been grasped within the formal classroom setting. Other students, while they grasp the concepts, will expand their understanding of statistical concepts through the use of projects. Learning and motivation for learning have been significantly increased in my classroom as a result of using supporting projects.

Many of the ideas for these projects have been developed through the sharing of ideas with fellow teachers. I thank each of them for giving their ideas to me, as I likewise shared mine with them. Time has erased the original forms of most of these activities, and I find with each new semester that I continue to modify them to fit the needs of each new class of students. Calculators have changed most of the random experiments and provided learning of new methods for random simulations. Computers and the Internet continue to provide resources for data. Students must use these tools in a hands-on environment where they can explore and seek solutions to open-ended questions in order to be prepared for the world they will enter into beyond their education.

The projects in this supplement have been organized to correspond with the topics covered in David Moore's statistics textbooks. I have included a section titled "Notes to Instructors" where the purpose of each project is listed along with any specific items needed to do the project. While many items can be used to generate data, I have found students have an increased motivation for learning by using food items as data collection devices. Buttons may come in all sizes and colors, but, in my classroom, M&M's are more fun, edible, and get the learning task accomplished.

Possible times to do these activities within the class are: a) within each chapter as the topic is discussed, b) at the end of the term for review, c) on shortened class weeks with out time to proceed with an additional unit, and d) as individual projects outside of the class to then be presented to the class in poster sessions or displayed on bulletin boards.

Enjoy the teaching and the students will learn. Enthusiasm is contagious and I make every attempt to spread it. Happy projecting!

January 2003

Notes to Instructors

The following summary provides a statement of purpose for each project and the materials needed.

Project 1 WHAT'S THE COST OF LUNCH?

Purpose:
To gather data and display it using both a graphical and a numeric method.

Materials needed:
None.

Project 2 SPRING BREAK

Purpose:
To look at the relationship of distance traveled and air fares. The student will use the Internet to find the lowest fares for a set of destinations for typical spring break trips and the distance of the flight. Possible Web sites are Map quest, Travelocity, and the airline Web sites, but others may be used at the teacher's discretion. A scatterplot of the data, a least-squares regression line, and correlation will be found. The student will give a persuasive argument for his or her conclusion.

Materials needed:
Access to the Internet.

Project 3 TEST YOUR MEMORY

Purpose:
To work with residuals. Choose a list of people whose names your students recognize. It adds interest for the students to include you in the list. You can find the birth dates of celebrities on the Internet by searching, for example, "Whoopi Goldberg birth date." I am listing a few favorites here:

Drew Bledsoe	02 – 14 – 72
Doug Flutie	10 – 23 – 63
Catherine Zeta–Jones	09 – 25 – 69
Drew Barrymore	02 – 22 – 75
Julia Roberts	10 – 28 – 67
Richard Dreyfus	10 – 29 – 47
Nick Carter (Back Street Boys)	01 – 28 – 80
Whoopi Goldberg	11 – 13 – 49
Ronald Regan	02 – 15 – 11

George W. Bush   07 – 06 – 46
Al Gore   03 – 31 – 48

I suggest that you use at least 10 people with a wide range in ages. Include yourself or some well-known member of the faculty for some class fun.

Materials needed:
People known to the students and their birth dates.

Project 4   POPULATION GROWTH

Purpose:
To use the Internet to find population data, and then use the data to make a prediction of the population for a chosen location in the next census.

Materials needed:
Access to the Internet.

Project 5   WHAT DO STUDENTS DRIVE?

Purpose:
To design a study and gather data.

Materials needed:
Access to a student parking lot.

Project 6   JELLYBEAN

Purpose:
To use capture-recapture random sampling to estimate the number of jellybeans in the jar.

Materials needed:
1 large bag of multicolored jellybeans (with no black)
1 small bag of black jellybeans
1 large clear jar or fish bowl
Utensils for stirring and dipping
Paper cups or napkins

To start, place the multicolored jellybeans in the bowl.

Project 7   FAIR COIN

Purpose:
To observe probability of a repeated event over time. The project may be done individually or as a group where each member of the group uses a different coin.

Materials needed:
Coins for each student.

Project 8   ODD AND EVEN

Purpose:
To explore binomial probability.

Materials needed:
One die to roll
One penny to move as a game piece

An alternative procedure is to use a random number generator on your calculator in place of the die.

Project 9   THE AGE OF A PENNY

Purpose:
To determine the approximate age of pennies in circulation. The distribution of the ages, not the year the coin was minted, will be examined. The central limit theorem will be used to form an interval estimation of the ages for your sample of coins.

Materials needed:
One roll of pennies for each student.

Project 10   AHOY MATES

Purpose:
To develop a sampling procedure, generate data, display the data with an appropriate graph, and perform a hypothesis test of the companies claim. The student is to know procedures for designing an experiment, using a randomizing process to gather data, and conducting a hypothesis test of means.

Materials needed:
One bag of CHIPS AHOY cookies with the label reading "over 1000 chips."

Project 11   SNAP CRACKLE POP

Purpose:
To compare two distributions using a hypothesis test for the difference of two means, using the P-value approach. This project may be done in groups to access more data in a timely manner.

Materials needed:
None, however the students will need to go out of the classroom to gather their data.

Project 12   TASTE THE DIFFERENCE

Purpose:
To model a taste test to the binomial distribution and perform a hypothesis test for proportions. The student is to gather data and use the binomial probability density function and model the binomial distribution with the standard normal distribution. The TI-83 calculator (or equal), the binomialpdf, and binomialcdf functions will be used.

Materials needed:
One one-pound bag of plain M&M's
TI-83 calculators

Project 13   PLAIN AND PEANUT

Purpose:
To analyze more than one data set and the differences in two or more distributions. The project may be used as separate parts for the comparison, or be used with the Chi Square distribution.

Materials needed:
One one-pound bag of plain M&M's
One half-pound bag of plain M&M's
One package of snack size plain M&M's
One one-pound bag of peanut M&M's
One half-pound bag of peanut M&M's
One package of snack size peanut M&M's
Access to the Internet
TI-83 calculators

Project 14 SHORT OR TALL

Purpose:
To find a line of best fit to a paired set of data values, use inferential methods of the linear model, and make prediction intervals.

Materials needed:
None

Project 15 MANY VARIABLES DO PREDICT

Purpose:
To use a computer to find multiple variable equations.

Material needed:
Access to a computer with Excel or Minitab.

Project 16 GROWTH GROUPS

Purpose:
To compare multiple groups using ANOVA.

Materials needed:
Access to a computer with Excel or Minitab.

Project 17 RANDOM ASSIGNMENT

Purpose:
To compare multiple groups using ANOVA.

Materials needed:
Access to a computer with Excel or Minitab
Class rosters with names

Project 18 SKITTLES MACHINES

Purpose:
Check data with statistical process control.

Materials needed:
Fun size bags of Skittles
TI-83 calculator

Project 19   NBA

Purpose:
Check data with statistical process control.

Materials needed:
Access to NBA data (Web site or newspaper)
TI-83 calculator

Project Final

Purpose:
To use five aspects of a statistical study. The project should be three to five pages in length. The student may use either an observational study or a designed experiment. This project is best used as the culminating activity for the class in order to demonstrate overall conceptual learning.

As instructor, approve the students topics before he or she begins the project to assure the appropriateness of the research topic.

Materials needed:
None

Looking at Data–Distributions

Project 1   WHAT'S THE COST OF LUNCH?

How much does a student spend on lunch during school days? The student may choose to go to the local franchise of "Betsy's Burgers," go through the a la carte line, or eat the Type A regular lunch. Any way you slice it, you still are paying for your lunch.

1. You are to gather data from 20 males and 20 females today at lunch and record the data below. Now remember this is not a scientific study, so you can take their word for what they spent for lunch. Record the data and indicate if the lunch was purchased on or off campus.

Student Count	Female On / Off	Amount $		Male On/ Off	Amount $
1					
2					
3					
4					
5					
6					
7					
8					
9					
10					
11					
12					
13					
14					
15					
16					
17					
18					
19					
20					

2. Make a stemplot of the 40 values in the space below:

3. Are there any outliers in the data set? If so, list them and give a reason to either eliminate them or keep them in the data set.

4. Find the five-number summary of the costs from the above data.

Low = ____ Q1 = ____ Median = ____ Q3 = ____ High = ____

5. Now let's look at the data as different sets. Make a back-to-back stemplot of the data by gender in the space below.

Female   Male

6. Make side-by-side boxplots of the Female and Male data sets in the space below.

7. How do the two distributions compare? Who spends more for lunch?

10

8. Now look at the data of the costs spent for lunch On / Off campus. Make a back-to-back stemplot of the data in the space below.

On Campus   Off Campus

9. Make side-by-side boxplots of the above data sets in the space below.

10. How do the two distributions compare? Where does the lunch cost more?

11. By looking at the stemplots and boxplots you have displayed in this project, which display gives you the clearer picture for the comparisons and why?

Project 2   SPRING BREAK

Spring Break is coming and it's time to make your plans.   Below is a list of possible destinations for your trip.   Give the following information to assist in planning your trip and finding the information needed for this project.

1. You are leaving from   _____

2. Date of departure   _____

3. Date of return   _____

You are to use the Internet to search for the lowest airfares from your town to the destination on the given dates and find the distance from your airport to that city.

4. Complete the following table of information for your Spring Break adventure.   The cost is for one person on a roundtrip ticket.   A maximum of two stops in route is allowed, for both directions of the trip.

List the Web sites that you use for your search:

DISTANCES   _____

AIRFARES   _____

Destination	Distance	Airline	Departing flight number	Roundtrip cost
Miami				
San Diego				
San Juan				
New York City				
Chicago				
Seattle				
Salt Lake City				
Boston				
Honolulu				
Denver				

5. In the space below, make a scatter plot of the distance versus cost. Use the distance as the independent variable on the horizontal axis. Make appropriate scales for the axes.

6. Look at the scatterplot above to answer the following: How is the cost of the trip associated with the distance of the trip?

7. Use a straight edge to approximate a line of best fit to the data. Approximate the slope of the line and the y intercept.   Write the equation of your estimated line, $Y = a + b X$

8. On a scale of 0 to 1, how good does the line fit the data? $0 =$ no fit and $1 =$ a perfect fit.

9. Enter the data for the distances and the costs in your calculator. Use L1 for the distances and L2 for the cost.   Press STAT, CALC, choose LinReg(a + bx) to find the equation of the least-squares regression line.

10. Find the value of the correlation coefficient for the data. The r value can be found as follows on the TI-83: press VARS, Statistics, EQ, r from the menu screen, or you may go to the catalog and select 'DiagnisticOn' then do the LinReg (a + bx) as indicated in question 9.

11. Sketch the least-squares regression line on the scatterplot above and compare it to the line that you guessed. Are they close to the same line?

12. In part 8 you guessed a value for the strength on the fit to your line. Compare it to the value of the correlation coefficient. How close were you?

13. By looking at the scatterplot and the regression line find a place that is a good buy for the miles traveled. Why do you think the airfare is below the expected value for this destination?

14. Use the destination identified in question 13 and any information found in this project to write a persuasive essay (50-100 words) convincing your parents to let you go to your destination of choice for Spring Break.

Looking at Data–Relationships

Project 3   TEST YOUR MEMORY

1. How well do you know the ages of the stars and celebrities that you
    hear or see on a regular basis?   You are to complete the following
    table with what you believe to be each person's age. After we have
    completed the list, you will be given their actual ages.

Name	Your guess of the age	Actual age	Residual	Absolute value of residual	Residual squared
George W. Bush					
Catherine Zeta Jones					
Nick Carter					
Drew Bledsoe					
Whoopi Goldberg					
Al Gore					
Richard Dreyfus					
Ronald Reagan					
Doug Flutie					
Julia Roberts					
Drew Barrymore					
Mr(s). Math					

2.   Now that you have filled in your guess, your instructor will give you their actual ages.

3.   To see how you have done we will analyze the differences between their actual age and your guess.   In the space below draw a coordinate axis system and plot the ordered pairs (X , Y) where X is the actual age and Y is your guess.

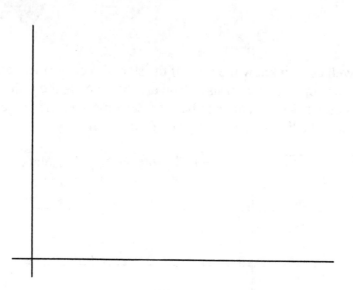

4.   Now sketch the line Y = X. If your points fall on the line, then you have correctly guessed the ages. If your point is not on the line, then you are to draw a vertical line from your point to the line. The distance of each of these segments is a residual for that guess. List these directed distances in the residual column next to the guess.

5.   Did you tend to underestimate, overestimate, or bounce around the line?

6.   Sum the residual column. Does your answer tell you how well you guessed the ages overall? Why or why not?

7. There are two choices to consider arriving at a meaningful sum for this data.

    a) Take the absolute value of each value and then find the sum.

    b) Square each residual and then find the sum of the squares.

8. You must decide as a class or a group which procedure above to use as the criteria for the next question.

9. Compare your results with the rest of the class and decide who has the "best" knowledge of the stars' ages.

10. Now use your TI-83 to do this problem. Use the STAT menu to enter the actual age of the star in L1 and your guess in L2.

    o In the STAT PLOT menu, make a scatterplot of the data using L1 as the Xlist and L2 as the Ylist. Choose ZOOM STAT to make a user-friendly window.

    o Use the Y= menu to enter the equation $Y1 = X$ and press the GRAPH key.

    o You should now "see" the graph that you made in the above space.

    o Return to the STAT EDIT menu and define L3 as $L2 - L1$.

    o Define L4 as ABS(L3).

    o Define L5 as $L3 \wedge 2$.

    o The LIST menu is found by pushing 2nd STAT. Now use the MATH menu to find:

    a) Sum(L4) = _____

    b) Sum (L5) = _____

11. Now that you have completed the above, write your definition of residual in the space below.

Looking at Data–Relationships

Project 4   POPULATION GROWTH

Choose a location that you would like to know the population. This may be a state, a county, a region of the country, or a city. Use the Internet to search the U.S. census reports.

1. List your choice in the space provided.

   _____

2. You are to research the population given by the U.S. census reports to find the population for each of the previous census reports from 1790-2000. List the populations in the table below:

Year	Population
1790	
1800	
1810	
1820	
1830	
1840	
1850	
1860	
1870	
1880	
1890	
1900	
1910	
1920	
1930	
1940	
1950	
1960	
1970	
1980	
1990	
2000	

3. Use your TI-83 to explore models for the growth of the population over time. Make a plot of the data by choosing STAT PLOT. Use Year as the X List and Population as the Y List for the data set. In the table below give the equation of four possible models of the data set and the value of R^2 for each model. Provide a sketch of the data and graph of the model by using the Graph Link or hand sketch for each of the four models. Attach the graphs to your report for this project.

Model	Equation	R^2

4. Now use the graphs with the supporting information to make a decision on which model best fits the data.

5. Give supporting reasons to your conclusion.

6. Using your model, predict the population in the next census.

Producing Data

Project 5   WHAT DO STUDENTS DRIVE?

What do the students drive to school? Your job, should you decide to accept it, is to design an experiment, gather data, and answer this question.

1. Adjacent to the school you have parking areas. You are to design a method that would produce a representative sample of 30 cars. You may use a block design, a stratified sample, a simple random sample, or other methods you choose. State your design in the space that follows.

2. List the location for each car that you have selected to include in your sample before you venture into the lot.

3. List the results of your sample in the space that follows. You are to identify one more variable to add to the list of characteristics that are given. Write it in the space provided as "other."

Car	Make	Year	Model	Color	Other
1					
2					
3					
4					
5					
6					
7					
8					
9					
10					
11					
12					
13					
14					
15					
16					
17					
18					
19					
20					
21					
22					
23					
24					
25					
26					
27					
28					
29					
30					

4. Summarize the data from your sample to describe a typical car in the lot.

Producing Data

Project 6   JELLYBEAN

How many jellybeans are in the bowl?

1. Make a guess of the number of jellybeans in the bowl and write your guess here.

2. Now it's time to do some "fishing". Each person in the room may use a utensil (the fishing pole) provided by the teacher to remove some jellybeans from the jar. (The total amount removed cannot exceed the number of black jellybeans that you have available.)

3. Replace the jellybeans that were removed with the same number of black jellybeans. (This is similar to the procedure of Wildlife Conservation Agents when the capture, tag, and release wildlife.) Write the number of replacement here.

4. Stir the jellybeans to mix in the black ones.

5. Use the fishing device to remove a sample of the jellybeans from the bowl as in an SRS.

6. Count the jellybeans that you have captured.

   # in total _____   # of black _____

7. Return your sample to the jar and have each person or group repeat the same sampling method.

8. Write a proportion and solve for the total number of jellybeans in the jar.

9. Now let's see how close we are with our prediction. Everyone participate and divide up the total number of jellybeans and count them! Write down the total number of jellybeans here.

10. Having been through this process, use your knowledge to answer the following question. The Missouri Conservation Commission is to determine the size of the Canadian Geese population in the state of Missouri. You are to design a capture-recapture procedure they could use to estimate the size of the flock.

Probability: The Study of Randomness

Project 7   FAIR COIN

Let's flip a coin and record the outcome. As a group, discuss the following items before you begin the experiment:
- How to flip the coin so it is done in the same manner each time.

- Whether or not to turn the coin when you catch it.

- Which side of the coin should start facing up? Alternate heads or tails up, or always the same?

- What do you do if you drop the coin . . . will you count the toss or re-flip?

1. You are to flip your chosen coin according to the agreed upon guidelines above. After each flip you are to record the results as an H (head) or T (tail). Keep track of the total number of heads after each flip and find the percentage of heads after each flip.

Record the results on the grid that follows:

Trial number	Start up H or T	Outcome H or T	Total number heads	% heads
1				
2				
3				
4				
5				
6				
7				
8				
9				
10				
11				
12				
13				
14				
15				
16				
17				
18				
19				
20				
21				
22				
23				
24				
25				

2. The above data gives you a numerical representation of the behavior of your coin. Make an observation about the percent heads column.

3. Let X represent the trial number and Y the percent of heads after X trials. Plot the set of (X, Y) on the grid below.

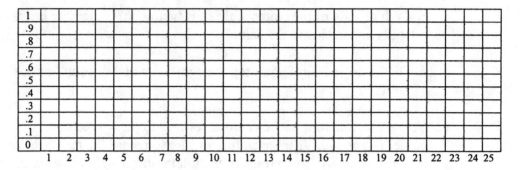

4. Make a line plot of the above points on the grid by connecting the points as you move from left to right on the grid.

5. The graphical display of the data on the grid gives you a picture of the behavior of your coin over time. What does your graph indicate to you about your coin?

6. By using both the numeric display and the graph, make a conjecture about your coin.

7. If the other members of the group used different coins, compare the graphs of the various coins. State your observations here.

   Penny

   Nickel

   Dime

   Quarter

8. Using what you have learned, consider the following. To start a volleyball match, the official begins with a coin flip. Do you think that you can call it correctly more than 50% of the time if you see the top side of the coin and know that she will not turn it over? Review your results of the coin flip column 2 and column 3. State what you observe.

9. Is this a fair method of starting the match?

From Probability to Inference

Project 8   ODD AND EVEN

Part 1:
The following grid consists of five rows and nine columns. You are to place a penny on the square marked START. With each throw of the die you will move down one row and left one square if the roll is odd or down one row and right one square if the roll is even. One game consists of four rolls of the die. The game will place your penny at some position on the bottom row. You are to play the game 16 times. Place a tally mark on the square in the bottom row position after each game.

But wait! Before you begin playing, you are to predict the final outcome of the 16 games by placing numbers in the bottom row that you anticipate as your final result. Yes, the sum of your guesses in the bottom row is 16.

Now you may play the game. "Let the good times roll!"

				Start				

1. So, how did you do with your predictions? Were you surprised at your results? Discuss symmetry and the cell locations that are impossible.

2. The model used in this game is called the binomial distribution. List the characteristics of a binomial model in the space below.

3. Use the binomial model to find the probability of ending on each of the five possible final positions of the above game board.

   You may use the binomialpdf function on the TI-83 to do your work, or the binomial theorem. Number the final positions 1, 2, 3, 4, 5 from left to right on the bottom row, and give the probability of ending in that position.

   Position 1 _____

   Position 2 _____

   Position 3 _____

   Position 4 _____

   Position 5 _____

4. Compare the results of the game to the result using the probability as computed above. List your observations in the space provided.

Part 2:
Repeat the game with the following changes in the movement rule. Roll a die and move down one row and left one space if the roll is 1 or 2, or move down one row and right one space if the roll is 3, 4, 5, or 6. List your results here.

Position 1 _____

Position 2 _____

Position 3 _____

Position 4 _____

Position 5 _____

1. How are the two games different?

2. How are the two games similar?

From Probability to Inference

Project 9   THE AGE OF A PENNY

Have you ever wondered how long coins stay in circulation? Are you a collector? You each have a tube of pennies. Your first task is to form a distribution of their ages.

1. Organize the data by using a stemplot of the ages. Split the stems to give sufficient stems to the data.

2. What is the shape of the distribution? Why do you think it is this shape?

3. Did you find any outliers?

4. Do you think the distribution of all pennies in circulation is similar to your sample?

5. List the characteristic assumptions for the Central Limit Theorem, and decide if they are satisfied by your distribution.

6. Find the mean and standard deviation of the ages of the pennies in your sample.

   Mean = _____   S.D. = _____   n = _____

7. Compute a 95% confidence interval for the mean ages of pennies.

   _____ < μ < _____

8. What is the margin of error for your estimate?

   M.E. = _____

9. The president of "COINS UNLIMITED" has just hired you as his chief statistician for his research on the age of pennies.   You are charged with the task of estimating the average age of pennies in circulation within one year of age with 99% confidence. How large of a sample would you need to obtain?   Use the standard deviation from your sample as your best estimate of the population standard deviation.

10. Consider your roll of pennies as a population and place a scale of the ages on the number line below. Choose 20 pennies at random from your pile of pennies. Find the mean and standard deviation of the sample and compute a 95% confidence interval for the population mean, μ. Draw a line segment for this interval below the number line that you have scaled.   Mix up the pennies and repeat the process five times.   Do the intervals of your sample capture the value of μ? Why or why not?

   ____/____/____/____/____/____μ____/____/____/____/____/____

11. On the basis of your research with this project, how would you define the age of a "rare" coin? Give a statistical definition for your choice.

12. What would be the age of the pennies that you would begin to save before they become hard to find? Give a statistical reason for your choice. Consider "rare" as 2% or less of the population.

Introduction to Inference

Project 10   AHOY MATES

The company claims, "There are over 1000 chips in every bag." How can they make such a claim? As a consumer, your job is to design an experiment to check the claim made by the company.

1. Describe a sampling procedure you would use to gather the data. Include how to select: the bags, cookies in the bag, and chips within the cookie.

2. Consider the sample size you think is needed for the problem. Choose a sample size for the number of cookies that will allow for an approximately normal distribution of the sample mean number of chips within a cookie.

3. You are now ready to begin your count. Most students consider this to be a "crummy" problem. State how you decided to "count" the partial chips?

4. Make a graphical display of the distribution of chips within the cookies.

5. Write a 95% confidence interval for the number of chips per cookie?

6. To consider the number of chips in a bag, you must now consider the number of cookies in the bag. While we have not chosen a large number of bags of these cookies, we can make an observation about the number of cookies in the bag. What is it?

7. Using the distribution of chips per cookie, and the number of cookies in the bag as your guide, write a 95% confidence interval for the number of chips in the bag by making a linear transformation of the data. $\mu_{\text{chips in bag}} = n\,\mu_{\text{chips in cookie}}$ and $\sigma_{\text{chips in bag}} = n\,\sigma_{\text{chips in cookie}}$

8. What do you think about the company's claim?

9. The company has just hired YOU as chief statistician. Address the problem with a hypothesis testing procedure. State the hypotheses (Ho and Ha) for the claim made by the company. Report the value of the test statistic and the p-value of the test.

10. Write a word conclusion (approximately 50 words) to the problem that you would submit to the company president.

11. Identify five items relating to statistics that you have learned from this project.

    1)

    2)

    3)

    4)

    5)

You may now eat the data. Got milk?

Project 11 SNAP CRACKLE POP

Are all cereals created equal? You are to gather data on different brands and types of cereals.

1. Go to the grocery store with pen and paper in hand, and gather some data on rice and oat cereals. Classify the cereal as to the first ingredient listed if it is made from a combination of grains. Record the data on a per-serving basis as follows:

Brand	Type	Calories	Total fat (g)	Sodium (mg)
	Rice			
	Rice			
	Rice			
	Rice			
	Rice			
	Rice			
	Rice			
	Rice			
	Rice			
	Rice			
	Oat			
	Oat			
	Oat			
	Oat			
	Oat			
	Oat			
	Oat			
	Oat			
	Oat			
	Oat			

In the data that you have collected, assume that the data is normally distributed for all cereals.

2. Perform a hypothesis test for the difference of the two means for each of the following questions. List the Ho, Ha, give the observed value of the test statistic, p-value, and write a word conclusion.

   a) Is there a difference in the calorie count for the two types of cereals?

   b) Do the rice cereals have less fat grams per serving than the oat cereals?

   c) Do the rice cereals have more sodium (mg) per serving than the oat cereals?

Inference for Proportions

Project 12 TASTE THE DIFFERENCE

Can you really taste the difference in the brown M&M's?

1. You are to select a partner to perform the following experiment. One person will record, the other will taste. After completion, you will reverse roles and repeat the experiment.

2. Take a small number (approximately 5) of brown M&M's and train your taste buds by eating them slowly one at a time.

3. Now take a non-brown M&M and eat it slowly. Sorry, no more than five tries of knowing each color, brown vs. non-brown. After all, if you eat all the data, we can't run the taste test!

You must trust your partner on this one to record the data, as you are the lucky one that gets to eat your way through.

4. Close your eyes; blindfolding is necessary to prevent looking at the candies.

5. The person recording should choose 10 M&M's of random colors. Make sure that there is a good color mixture, but no more than five brown. One by one, the taster should eat the candies given to you by your partner, and tell him/her if you think it is brown or not brown. The partner is to record the results as correct or incorrect for each trial. Tell the taster if they are correct after each trial. Record the results for each of the 10 trials as T = true for a correct guess and F = false for an incorrect guess in the chart below.

Trial #	True or False
1	
2	
3	
4	
5	
6	
7	
8	
9	
10	

6. This experiment consists of independent trials, each of which has two possible outcomes, success or failure, where the probability of success and failure add to one. If a person has no idea as to the color, then the probability of success is .5 for every trial. Count the number of successes that were observed in the 10 trials above. Find the probability of that number of successes if indeed you were guessing. You may use the binomial theorem or use the binomialpdf function on the TI-83 calculator. Express the method of your computation in the space below.

7. Make a probability density function for the random variable of the number of successful guesses in the 10 trials. Use $p = .5$ and list the results in the table below.

Number	Probability
0	
1	
2	
3	
4	
5	
6	
7	
8	
9	
10	

8. Find the expected value of the number of successes for the distribution from the above.

9. Find the standard deviation for the above distribution.

10. Let n equal the number of successes that you observed in your taste test. By using the standard normal distribution, find the $P(X \geq n)$. Use the continuity correction factor.

11. On the TI 83, choose the binomialcdf function and find the $P(X \geq n)$.

12. Compare your answers to question 10 and question 11. Are they the same or different? Why?

13. So how did you do? Do you think now that you really can tell the difference? Are you willing to bet your leftover M&M's?

14. Switch sides of the table and repeat the taste test with your partner. Record the results here.

15. So who did the better tasting? _____ Repeat the experiment so that each of you have 20 trials. Let k equal the total successes in the 20 trials, and find $P(X \geq k)$ by using the standard normal distribution with $n = 20$ and $p = .5$.

16. Test the hypothesis that you are guessing vs. the alternative that you can tell the difference. Use the p-value techniques to test your hypothesis and write a word conclusion that could be used as evidence in court to support your findings.

Now eat the evidence so that the next hour class is unaware of "how sweet it is" in here!

Inference for Two-Way Tables

Project 13   PLAIN AND PEANUT

What is your favorite color of M&M? Can you really taste the difference? When you buy the bag do you eat them first or last? The real question here is to address the proportions of colors of M&M candies. Over time, the company has changed the colors and proportions.   So, what are the proportions now? You will discover the answers to these and other questions as you proceed with this project.

1. Identify the colors of plain M&M's that are in the package? Name the colors.

2. Before you begin the data collection, state the percentages of colors that you *think* are in the bag.

3. Discuss with the class and determine a procedure for counting the M&M's. Consider the bag sizes in making your choice and the size of the sample. Identify your procedure here.

4. Make a frequency distribution of the counts and give the relative frequency for each.

5. The "color" of M&M is what type of data? _____

6. Graph the data by using an appropriate chart. State why you choose your type of graph.

7. Repeat the procedure using peanut M&M's. What are the colors used by the company?

8. Make frequency distributions of the counts and find the relative frequencies.

9. Graph the data using the same type of graph used in the plain M&M data.

10. Compare and contrast the two distributions and graphs.

11. How do we know what the company intends to be the proportions for each color? Consult the Web site for the MARS Company.

12. List the findings from the company Web site for both the plain and peanut M&M's.

    a) Plain
       List the colors and the company's proportion claim.

    b) Peanut
       List the colors and the company's proportion claim.

13. Does the data that you have gathered agree with the company's claims?

14. Identify a procedure that you could use to check the goodness of fits for your data.

15. Identify the Ho and Ha you would use to verify the company's claim.

    Ho:

    Ha:

16. Display in a matrix the observed and expected cell values and perform a Chi-Square analysis for the (a) plain and (b) peanut data.

    a) Plain

Color	Observed	Expected	(O-E)^2 / E

b) Peanut

Color	Observed	Expected	(O-E)^2 / E

17. Write a word conclusion to the hypothesis test you have performed using the 5% level of significance.

a) Plain

b) Peanut

18. Now consider the type of M&M and the color. Use the two samples that you counted to test the hypothesis that the type of M&M is independent of the colors as listed. State the hypotheses, the value of the test statistic, the p-value, and write a word conclusion.

Use the following grid cells to display your data using the columns as types and the rows as colors.

	Plain	Peanut	Total
Red			
Green			
Yellow			
Blue			
Purple			
Total			

Inference for Regression

Project 14   SHORT OR TALL

Do tall people have big feet and short people have small feet?   Your task is to generate a set of data values for the height in inches and their shoe size.

1. In the chart below list at least 15 people and their data. You are to standardize the sizes for male and female.

Name	Height	Shoe size

2. Make a scatterplot of the data where shoe size depends upon the height.

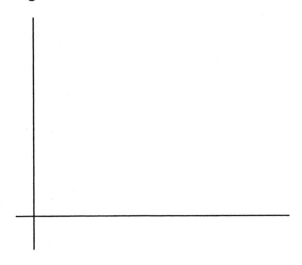

3. Find the line of best fit to the data and write the equation.

4. Make a residual plot of the data here.

5. What does your residual plot indicate to you about the model?

6. Perform a hypothesis test to determine if there is a significant positive slope to the line of best fit. You may use the LINREG t test menu on your TI-83.

   o Write both hypotheses.

   o State the t-value.

   o State the P-value for this test.

   o Write a word conclusion.

7. Suppose Sue Z. Normal walks into your classroom. Find a point estimate for the shoe size of Sue Z. Normal who is 70 inches tall.

8. Find a 95% confidence interval for the shoe size of Sue Z. Normal.

9. You may not have had anyone in your sample that was over seven feet tall, but there are seven footers around, particularly on pro basketball teams. Predict the shoe size for Mr. Guy Tall, who measured in at 7 feet 2 inches for Coach Woy Rilliams at the basketball tryouts.

10. Find a 95% prediction interval for Mr. Guy Tall's shoe size.

11. State why the prediction interval for Mr. Tall is wider than the confidence interval for Sue Z. Normal.

Multiple Regression

Project 15   MANY VARIABLES DO PREDICT

Can you predict your arrival time at school? While there are many variables that could impact this time, we will consider only the two variables of the distance that you are from school and the time that you leave home to go directly to school.

1. Gather information using the time in minutes starting at $t = 0$ as 7:00 am and give the distance in tenths of a mile. Each member of the class is to record the data and return it tomorrow for class.

Record the information in the table. Use a minimum of 25 sets of values.

Arrival time	Distance	Start time

2. Let Y = the arrival time, $X1$ = the distance, and $X2$ = the start time to find a multiple regression equation for the relationship. You are to use a spreadsheet to enter the data and generate a multiple regression output for the variables defined above. Excel or Minitab will output this information. Attach a copy of the output to this report.

3. Write the estimated equation from the output.

4. State the hypothesis tested by the ANOVA F statistic for this problem in words.

5. State your conclusion to the hypothesis based on the computer output.

6. What percent of the variation in arrival time is explained by the distance and start time?

7. If you live 5.3 miles from school and leave at 7:15 am, what is your expected arrival time?

One-Way Analysis of Variance

Project 16   GROWTH GROUPS

As you wander the hallways of the school, you note various heights of students. For this project you are to select a random sample by means of a convenient method to compare the four grade levels of students and their heights.

1. List the heights of a minimum of 15 students of the same gender from each of the four grade levels in the space provided.

Freshman	Sophomore	Junior	Senior

2. Make side-by-side boxplots for the data sets listed above.

3. Summarize the relationship of the boxplots. How are they alike and how are they different?

4. Make a normal quantile plot for the data of each of the four grades, and state your conclusion regarding the normality of each.

    a) Freshman

    b) Sophomore

    c) Junior

    d) Senior

5. Summarize the data for each class in the table below.

	Freshman	Sophomore	Junior	Senior
Mean				
Standard deviation				
Sample size				
Standard error				

6. Assume that the groups have equal standard deviation and perform the one-way ANOVA test. State the hypotheses.

7. State the degrees of freedom for the ANOVA test.

8. Summarize the results in the ANOVA table below.

Source	Degrees of freedom	Sum of squares	Mean sum of squares	F
Groups				
Error				
Total				

9. You may use a computer output to do the computation for this problem and attach a copy of its output to this project.

10. State the critical value of the test statistic. Use an α level of 5%.

11. Summarize the conclusion to the hypothesis test. Include the P-value and state your results in terms of the variables in the problem.

ANOVA

Project 17   RANDOM ASSIGNMENT

In a large school where the students are assigned to their class schedule by computer, it would sound reasonable that the students within all classes of the same topic for example, Algebra II would have a similar distribution of last names.

Your instructor has noted that in each of his classes this year the students seem to be clustered in an alphabetical grouping. Morning classes tend to have the students with last names like Adams, Beltron, and Jones, but the afternoon classes seem to have Smith, Thompson, and Young.

Is this true in your classes? Choose an instructor and make a distribution of the first letter of the last name for each member of the class. You are to use three classes for this project. Recall that we are checking for the assignment of students within the subject matter, so more than one section of a class must exist. Assign a numeric value to each letter of the alphabet as follows: a = 1, b = 2, c = 3, etc.

While the distribution of values will not be normal, the central limit theorem indicates that the sample means will tend to approach a normal distribution as the sample size approaches 30. Use this fact to decide if the students in the classes could be considered a random assignment to the daily class schedule.

1. Make a stemplot of the data for each class in the space below.

2. What relationships, similar or different, do you see in the graphs?

3. Make side-by-side boxplots for the above data in the space below.

4. Compare the plots and state you conclusions.

5. Perform a hypothesis test by comparing the means of the classes using a one way ANOVA on the three distributions. Write the Ho, Ha, give the value of the F statistic, and state the P-value for this test.

6. Write a word conclusion that you could present to the principal of your school regarding the random assignment of students to the various periods of the daily schedule.

Statistics for Quality

Project 18   SKITTLES MACHINES

You have just been hired as the quality control engineer for Skittles. Your job, because you decided to accept it, is to watch the process of the bagging of the fun size bags to insure that they are meeting the company's quality standards is the following ways:

A. Are the number of candies in the bag within the company's guidelines?

B. Are the colors of each candy within the proportion guidelines?

Assume that your bags of fun size Skittles are being selected randomly from the bagging machine at one hour intervals. Select the bags one at a time and count the data to assess if the machine is functioning properly. You are to consider the following three methods of control:

    i. Any point outside three standard deviations from the mean.

    ii. Any nine consecutive values on the same side of the mean.

    iii. Any two of three points more than two standard deviations on the same side from the mean.

1. In the space below make a quality control chart for the number of Skittles in the bag. Company standards are for the bags to have a normal distribution with mean 18.2 and standard deviation 0.6. Experience has shown that the machine produces bags with equal proportion of the five colors. Label the control values of the vertical axis. Plot the number of candies in each bag, in order selected, on the horizontal scale.

1 2 3 4 5 6 7 8 9 10 11 12 13 14
Bag

Use the above chart to determine if the process is in control or indicate the point where you have determined it to be out of control.

2. The bagging machine is fed by machines that produce each color of the candy. You wish to determine if each of them is remaining in control throughout the day. You are to make a control chart for the proportion of each color. Assume that the colors are of equal proportion (p = .2). Make the control limits for the proportion and then plot the proportions from each bag by color, as you sample.

Color RED

1 2 3 4 5 6 7 8 9 10 11 12 13 14
Bag

Color Purple

1 2 3 4 5 6 7 8 9 10 11 12 13 14
Bag

Color Orange

1 2 3 4 5 6 7 8 9 10 11 12 13 14
Bag

Color Yellow

1 2 3 4 5 6 7 8 9 10 11 12 13 14
Bag

Color Green

1 2 3 4 5 6 7 8 9 10 11 12 13 14
Bag

3. Write your conclusions about the performance quality of the bagging machine and each of the color production machines in the space below as a report to your boss on the day's activity.

A) Bagging machine

B) Red

C) Purple

D) Orange

E) Yellow

G) Green

Project 19   NBA

You are to select a team from the NBA to do the following study by making a control chart for your team. As an avid fan of basketball, you have decided to chart the points scored by your team, the
_____. You may use the Web site www.nba.com to find the data for this project as a retrospective study, or chart the next 15 games that your team plays by checking the papers on a daily basis.

Past experience has shown the NBA teams average 98 points per game with a standard deviation of 3 points per game. You expect your team to be consistent. Consider the following three measures of them experiencing a significant change:
  i. Any game outside three standard deviations of the mean.
  ii. Any nine in a row on the same side of the mean.
  iii. Any two of three more than two standard deviations on the same side of the mean.

Team _____

```

```

1 2 3 4 5 6 7 8 9 10 11 12 13 14 15
Game

List the date for each game for reference.

Write a summary of your team's performance over the past 15 games. Note any trends that appear on your graph. If anything happened to any player on the team during the time or if any other significant event(s) took place, which could have accounted for the change in team performance, please reference them in your report.

Research Project

Project Final

Within this project you must demonstrate five different aspects of a statistical study. The project should be three to five pages in length. You may use either an observational study or a designed experiment. Your instructor must approve your topic of research before you begin.

1. Pose a question that you can answer with data (this must be pre-approved).

2. Explain how you will select the data or design the experiment to gather the data. Give a rationale for your choices.

3. List the data values you obtain and choose an appropriate graphical display for the data. The descriptive methods are to relate to the question that you have asked.

4. Use a hypothesis testing procedure to analyze the data. Give supporting reasons for the choice of the procedure used. Summarize your conclusion in terms of the question that you posed. Include the p-value of the test.

5. The final paragraph should summarize what you have done in the form of a letter to your Grandma or Grandpa, using language that she or he will understand.

Statistics in Research

John C. Turner
U. S. Naval Academy

Preface to Statistics in Research

One of the current ideas in statistical education is that courses should use "real world" data, wherever possible. Many texts do this by citing data from various studies. This is fine, as far as it goes, but this approach has several shortcomings. For one thing, the student gets an extremely brief view of the scientific problem. Further, it is a highly "sanitized" view of the problem. The textbook author has read the article and extracted precisely the information that is needed to do the problem discussed in the text. At the U.S. Naval Academy, I have put together a course for students who have completed introductory statistics. In this course, the students read actual journal articles and discuss how statistics was used in the conclusions of the article. This led to a number of observations. Sometimes, necessary data (such as sample size) is either buried elsewhere in the article or completely missing. Some articles used inappropriate methods or mis-interpreted the results of the analysis. This includes, for example, confusing small p-values with large effects. Lastly, by considering the entire article, the students got a much better idea of how the statistical analysis related to the scientific conclusions of the article.

Since starting the course described above, I have also used articles in basic statistics, notably at The University of Texas at Austin. The students were quite good at following the statistical analysis and some could even follow analyses that had not been covered in the course. Many of the students commented that they were impressed that they could make immediate use of what they had learned, even though it was a very introductory course.

I have tried to include articles from a number of fields and I hope that the students will find several articles of interest. I have tried to use articles that do not require much specialized scientific knowledge to follow. Above all, it is my hope that by following the entire article, the student will come to appreciate that statistics is not just a bunch of formulas that you guess at on a test, but rather a method of analyzing data that is critical to much scientific research.

Introduction

One of the most interesting aspects of statistics is that it has application in a wide variety of problems in a wide range of fields. Any situation where we wish to use sampled data to draw broader inferences is fodder for statistics. In many situations, of course, considerable background knowledge is required in order to make good sense of the problem being solved and the significance of the solution. However, there are many problems that can be addressed without much in-depth knowledge of the particular field. I have collected several such papers for use as a companion in a beginning statistics course.

The term "beginning statistics course" encompasses a wide range of courses. There is not a single, well-defined set of topics that would be included in such a course. However, there *is* a set of topics that surely would be covered by any such course and also a slightly larger set of topics, only some of which would be included in any particular course. There is also not a set sequence of these topics. The placement of such topics as contingency tables and ANOVA is somewhat variable, if they are included at all.

This leads to several issues in the use of this material. There are two broad approaches available. One is to utilize all (or many) of the articles as progress is made through the course. When covering the different types of variables, for example, the student could consider all of the articles and identify the types of variables in each case. Then, when experimental design is covered, the student could return to that set of articles and discuss the design in each one. This could continue throughout the course. Of course, when the course covers a specific statistical method, only the papers that use that method would be considered.

Alternatively, use of the articles could be postponed until the first statistical tests have been covered. Then, the student might consider some of the articles that use the methods studied thus far. For these articles, the student could then identify all the variables in these articles, as well as discuss the sampling scheme and experimental design, etc. With this approach, when a new method is covered, the student would repeat the exercise with an article that makes use of the new method.

A third approach would be a combination of the two above. Early in the course, before statistical methods have been covered, the student would use only some of the articles and discuss variables, sampling, and design. Then, when a particular method is covered, the student can perform the full analysis of a paper that uses the method just covered. If this article is one that has previously been discussed, then the student will have already covered the notion of variables and design. If the appropriate paper for this method has not yet been covered, then the student will begin with the identification of variables, etc.

Reaching over all these concerns is the interests of the students. I have tried to include a variety of fields, for several reasons. A given class will find some articles more familiar or more interesting than others. The instructor will do well to bear this in mind. At the same time, it is possible to make use of articles outside of those that are presented here. The instructor should take particular care in selecting additional articles to see that the articles are appropriate both in the scientific problem being addressed and in the statistical approach. There is also a decision in terms of whether to use articles that illustrate poor uses of statistics. There are many such examples, but I have chosen not to include any in this supplement. Nonetheless, the students may learn quite a bit from a poor application of statistics, if the example is carefully chosen and it is made clear that the point is for the student to learn what to avoid, rather than simple criticism of the article.

Common Tasks for All Articles

1. Read the article. Some of the terms may be unfamiliar, but you will have to decide which terms are crucial and which are not. For instance, it is enough to know that an article deals with a specific species of fish. You don't need to know that many details about that species. It is probably a good idea to read the article all the way through first. Then review the general flow of the article. Some articles start with one problem, but then discover something new partway through. Try to see all the parts and see how they fit together. In many experimental papers, the first section will demonstrate that there are no significant differences between the treatment and the control groups on important variables. Thus, if differences are noted after the treatment, it is logical to consider them as resulting from the treatment, rather than from some initial differences between the groups.

2. Determine the variables in the study. There may be quite a few, especially in cases where the article has multiple objectives. Which variables are quantitative and which are categorical? Some texts discuss a third type of variable – ordinal. These are like categorical in that you can't do arithmetic on them, but have the additional property that they do have a natural ordering. A common example is where responses to a survey are "Strongly disagree," "Disagree somewhat," etc. Data like this (called a Likert scale) are usually converted to 1-5 or 1-7 and then considered to be quantitative. It is less clear what to do with categories such as Young, Medium, and Old.

3. Very few journal articles have graphical displays of the data. You should consider what is lost by not having these. Graphical displays are very helpful in giving an overview of the data to start with. In some types of analysis, the assumption is that the data have a normal distribution, at least approximately. A histogram may suggest how good this assumption is. The student should also bear in mind that a journal article does not show everything. Except for a few rare cases, you will not see all the data. You can at least hope that someone looked at the appropriate graphics before doing the analysis. Are there certain variables that you suspect are not normally distributed? How important is this to the results?

4. Sampling and experimental design are issues that are sometimes discussed explicitly, sometimes implicitly and sometimes not at all. If the article reports on an observational study, what "lurking variables" might be present? Did the authors make any note of them and try to correct for them? (You might not know how to correct for these other variables, but there are methods that can be used.) In a matched-pairs experimental design, why was the matching necessary? How would the authors go about doing the matching? Also, was the experiment "double blind?" It is not possible to make some experiments blind. How might this affect the results?

5. The instructor and the student alike need to be aware that there is more than one way to achieve the same result. Many texts (correctly) discuss the Z-statistic for testing two proportions. This same (two-sided) test can be done with a χ^2 statistic with 1 df. Similarly, the two-sample t test can also be done using an F statistic with 1 df in the numerator, if you consider the two-sample t test as a simple ANOVA. The (two-sided)

p-values are exactly the same for each of the equivalent methods. For the student, the important thing is to consider what test *you* think should have been done. If enough data is given, you can reproduce the test and compare your p-value to the one given.

6. As you read through the analyses, consider the number of sides and the p-value. It is quite common to find authors doing two-sided tests when a one-sided test is more appropriate. What is the effect of using the wrong number of sides? Is this effect the same when the p-value is quite small as when it is more modest? It is common to find that articles only state that the p-value is less than 0.001, say, and not give the exact value. Why do they not give the exact value? Does this affect the usefulness of the article from the point of view of the intended audience?

7. You will also notice that some articles only test hypotheses, some only give confidence intervals and some do both. What are the advantages and disadvantages of each method (tests and intervals)? In a given setting, does one approach give a better (more useful) answer than the other? Are there problems where it is not feasible to give a confidence interval?

8. Did the article stop short of truly answering their question? In ANOVA and contingency tables, we can test if all groups are equal. If they are not all equal, it would seem reasonable to consider which groups are different from which. This is not a simple matter, but would be a good approach to the question of interest.

9. It is important to repeat the calculations in the article, whenever possible. This makes it clearer what the authors actually did. It can also point out typos (which are more common than you might think). Sometimes, it also can clarify what calculation the authors actually did. The article might not be clear on whether the p-value is one- or two-sided. By performing the calculations, you can determine for yourself which one it is. Further, doing the calculations yourself helps you see what values are needed. In more than a few articles, the value of the sample size(s) can be hidden many pages away from the calculation or missing entirely. It is not uncommon to find that, while averages are given, the standard deviations are not, making it impossible to reproduce the calculation. On this point, note that some articles will give standard deviations and some will give standard errors, the standard deviation divided by the square root of the sample size. Sometimes, the terminology in the article is unclear and sometimes it is simply wrong. You will find out once you try to do the calculations yourself.

Learning What Not To Do

There are regrettably many examples of poor statistical practice to be found in journal articles. One of the purposes of publishing scientific articles is to spread information. There is not enough space in a journal article to cover every detail of the research. At the same time, it is important to give some support for the conclusions of the article. One part (and only one part) of this support comes from statistics. The student should consider the statistical evidence in a paper in this light – does it convince the reader that the conclusions of the paper are right?

Students often have trouble seeing this perspective when they read an article. However, one semester I had a student who was a female soccer player. She reported on an article that suggested a change in training methods to reduce ACL injuries (which are an increasing problem among female athletes). After she gave her report, I asked her, "Based on this article, would you change your training methods?" She thought for a moment and then gave a sincere explanation of why the evidence in the article was not convincing to her.

There are a variety of common omissions in journal articles and some are more serious than others. I have listed some of the possible omissions and a few ideas on how important each may be.

1. Failure to specify N. When N is omitted, the reader is completely at the mercy of the authors to interpret the statistics given. I would be very skeptical of a paper that would not include such basic information.

2. Failure to give standard deviations. When dealing with means, omitting standard deviations is the same as not giving the units of the data. I would again be skeptical.

3. Failing to distinguish between standard deviations and standard errors. Some authors appear not to know the difference between the two. Others will simply include one or the other and not be clear about which one is given. Sometimes, you can "back calculate" and determine which value was given. For instance, if a difference is claimed to be significant, you can calculate the t-statistic assuming the given value is a standard deviation and also assuming the value is a standard error. If one is significant and the other isn't, then you know whether the value is a standard deviation or a standard error. If they are both significant, then you are still lost.

4. Failure to graph the data. Research journals have very few graphs that look like ones in statistics texts. The text should make it clear that an important purpose of these graphs is in the initial phase of the analysis. We hope that the authors of the journal article did similar graphs when they began their analysis.

5. Do the data appear to be normal? This is similar to the previous point. We hope the authors did some consideration of this problem. Also, the student should remember that many of the methods (t tests in particular) are rather insensitive to deviations from normality. We generally need only that the distribution be mound-shaped.

6. ANOVA and t tests when the standard deviations are quite dissimilar. Most texts on ANOVA stress the need for at least similar standard deviations (within a factor of 2, typically). Many texts do not make a similar claim for the t test, even if the text points out how the t test is a special case of ANOVA. An article I chose not to include had done ANOVA where the largest standard deviation was over 1000 times larger than the smallest standard deviation!

7. Extremely large df in t tests. Most texts teach that the t distribution is very close to normal for df > 30. However, there are journal articles that report t-values with df in the thousands or tens of thousands. This is not incorrect, but is still disturbing. To me, it sounds like the author considers that statistics is a necessary evil for getting an article published, rather than a source of information about the data.

8. Simple typos. These are not serious and I have not seen any cases where typos have altered the conclusions of the paper. Further, any critic of typos is doomed to make one himself. They can be interesting for the student to catch and demonstrate that they know what some of the values should be. One example I have seen includes a list of t-statistics where two of the values were identical, but had different significance levels (for the same N). Enough other data were given to determine which value was right and which was wrong. The significance level given was correct and the typo turned out to be in the statistic itself.

The articles

1. "Antipsychotic Medication Adherence: Is There a Difference Between Typical and Atypical Agents?", Christian R. Dolder, PharmD, Jonathan P. Lacro, PharmD, Laura B. Dunn, MD, Dilip V. Jeste, MD, *American Journal of Psychiatry*, Vol 159, pp. 103-108.
2. "Does the Payment of Incentives Create Expectation Effects," Eleanor Singer, John Van Hoewyk, Mary P. Maher, Public Opinion Quarterly, Vol. 62, Issue 2, pp. 152-164.
3. "Efficacy of St. John's Wort Extract WD5570 in Major Depression: A Double-Blind, Placebo-Controlled Trial," Y. Lecrubier, MD, G. Clerc, MD, R. Didi, MD, M. Kieser, PhD, *American Journal of Psychiatry*, Vol. 159, pp. 1361-1366.
4. "Biology of a deep benthopelagic fish, roudi excolar, *Promethichthys promestheus* (Gempylidae), off the Canary Islands," Jose M. Lorenzo, Jose G. Pajeulo, Fishery Bulletin, 97(1) 1999, pp. 92-99.
5. "The Independence Gap and the Gender Gap," Barbara Norrander, *Public Opinion Quarterly*, Vol. 61, Issue 3, pp. 464-476.
6. "Moxibustion for Correction of Breech Presentation – A Randomized Controlled Trial," Francesco Cardini, MD, Huang Weixin, MD, *JAMA*, Vol. 280, No. 8.
7. "Undertreatment of Osteoporosis in Men with Hip Fracture," Gary Kiebzak, PhD, Garth A. Beinart, MD, Karen Perser, BA, Catherine G. Ambrose, PhD, Sherwin J. Siff, MD, Michael H. Heggeness, MD, PhD, *Archives of Internal Medicine* Vol. 162, No. 19.

"Antipsychotic Medication Adherence: Is There a Difference Between Typical and Atypical Agents," Christian R. Dolder, PharmD, Jonathan P. Lacro, PharmD, Laura B. Dunn, MD, Dilip V. Jeste, MD, *American Journal of Psychiatry*, Vol 159, pp. 103-108.

Background

Psychosis is a serious mental illness. It is best controlled with medication. However, standard ("typical") medications have serious side effects. On the other hand, if the patient stops taking medication, the illness can return. Psychosis can lead to a disconnect with reality, which makes it hard to get a patient who has relapsed to get back on medication. "Atypical" medications are ones that can be used to treat other conditions, but can also be useful for psychosis. They may have less severe side effects. In this article, the term "agents" means medications.

Questions

1. "Adherence Rates" are discussed on p. 105. What test was used to compare the 6 and 12 month rates? Does this test seem appropriate? In parentheses, the authors include the note "unequal variances." To what does this refer? What is the importance of this note? Do you think they needed the use the test based on unequal variances? Can you reproduce their calculations?

2. The authors use t tests for the quantitative variables, such as Age and Number of medications (Table 1). Does this seem like the most appropriate test? Should you be concerned about the relative sizes of the mean and SD for Number of Medications? What about the same issues for the cumulative mean gap ratio method?

3. The authors state that "significant differences among some of the five medications existed." What test did they use for this? Are they saying that all five medications are different?

4. The authors state that "olanzapine had a significantly lower gap ratio compared to haloperidol." What test did they perform? What are the special considerations needed?

5. Discuss the authors' conclusions for the compliant fill rate method. Can you confirm their calculations? What did they conclude about the 12 month fill rate? Does this mean there is or is not a difference at 12 months?

Antipsychotic Medication Adherence:
Is There a Difference Between
Typical and Atypical Agents?

Christian R. Dolder, Pharm.D.

Jonathan P. Lacro, Pharm.D.

Laura B. Dunn, M.D.

Dilip V. Jeste, M.D.

Objective: Pharmacy refill records were used to compare medication adherence in outpatient veterans receiving typical versus atypical antipsychotic medications.

Method: Consecutive patients meeting selection criteria and receiving prescriptions for haloperidol (N=57), perphenazine (N=60), risperidone (N=80), olanzapine (N=63), and quetiapine (N=28) over a 3-month period were identified from a computerized database. The hospital policy at the time of this study required failure in trials of at least two typical antipsychotics before initiation of an atypical agent. Patients' adherence with the antipsychotic regimen was calculated by analyzing refill records for up to 12 months. The cumulative mean gap ratio (the number of days when medication was unavailable in relation to the total number of days) and the compliant fill rate (the number of prescription fills indicating adherence in relation to the total number of prescription fills) at 6 and 12 months were calculated.

Results: Adherence rates at 6 and 12 months were moderately higher in patients who received atypical antipsychotics than in those who received typical agents. Cumulative mean gap ratios were 23.2% for typical and 14.1% for atypical antipsychotics at 12 months; thus, patients who received typical agents were without medication for an average of 7 days per month, compared with 4 days per month for those who received atypical agents. At 12 months, compliant fill rates were 50.1% for typical and 54.9% for atypical antipsychotics.

Conclusions: Interventions to improve adherence are warranted even for patients who receive atypical antipsychotic medications.

(Am J Psychiatry 2002; 159:103–108)

Effective treatment of psychosis involves the use of antipsychotic medication. Yet, reported rates of nonadherence (noncompliance) to antipsychotics range from 20%–89%, with an average rate of approximately 50% (1, 2). In patients with schizophrenia, nonadherence to antipsychotic maintenance treatment leads to psychotic relapse, rehospitalization, and more frequent clinic and emergency room visits (3, 4). The consequences of nonadherence thus contribute significantly to schizophrenia's estimated annual cost of $33–$65 billion (5, 6).

The improved side effect profile of atypical antipsychotics (i.e., lower incidences of extrapyramidal symptoms [7–10] and tardive dyskinesia [11–12], compared with the incidences for typical antipsychotics) has led investigators to speculate that patients receiving these medications will show greater adherence (13–15). We are aware, however, of only a few published reports comparing typical and atypical antipsychotics in terms of medication adherence. Rosenheck and colleagues (14) evaluated medication continuation and regimen adherence in 423 patients taking haloperidol or clozapine as part of a double-blind, randomized trial. Although the patients who received clozapine continued their medication significantly longer, the treatment groups did not differ in the proportion of pills re-

turned each week (14). Olfson and colleagues (16) examined the effect of antipsychotic type on adherence 3 months after 213 inpatients with schizophrenia or schizoaffective disorder were discharged while receiving typical (84.5% of patients) or atypical (14.5% of patients) antipsychotics. Patients were considered nonadherent if they reported stopping their antipsychotic for 1 week or more. A nonsignificant trend toward increased adherence was reported among patients with prescriptions for atypical antipsychotics (16). Cabeza and colleagues (17) retrospectively studied the relationship of adherence to antipsychotic type (typical agent, clozapine, or risperidone) in 60 inpatients with schizophrenia. Adherence over the past year was rated as adequate or irregular. No significant association was found between adherence and type of antipsychotic (17).

The purpose of the study reported here was to examine adherence to medication regimens of typical versus atypical antipsychotics among Department of Veterans Affairs (VA) outpatients with psychotic disorders. We selected pharmacy refill records to measure adherence because direct measurement of medication consumption in outpatients was not feasible. All of the methods commonly used to evaluate antipsychotic medication adherence in outpatients have limitations (18). Although pharmacy records provide an indirect measure of adherence, their validity for assessing adherence has been previously reported for nonpsychiatric and psychiatric outpatients receiving long-term pharmacotherapy (19).

We sought to test the hypothesis that patients with prescriptions for atypical antipsychotics (i.e., risperidone, olanzapine, or quetiapine) would be more adherent than those with prescriptions for conventional neuroleptics (i.e., haloperidol or perphenazine). We also examined the association of certain patient and treatment-related factors with nonadherence. On the basis of the available literature (20–31), we hypothesized that higher daily doses of antipsychotics and a larger number of adjuvant psychotropic medications would be associated with higher rates of nonadherence, but that age, gender, ethnicity, and diagnosis would not be associated with greater nonadherence.

Method

Patient Selection

This protocol was approved by the University of California, San Diego, Human Subjects Committee. By using the VA San Diego Healthcare System pharmacy computer database, we identified more than 1,800 new-fill or refill prescriptions for haloperidol, perphenazine, risperidone, olanzapine, and quetiapine within a 3-month time period. Haloperidol and perphenazine were chosen for this study because they were two of the most commonly prescribed typical antipsychotics in the VA San Diego Healthcare System. Patients with a prescription for clozapine were not included in this study because the VA San Diego Healthcare System required patients who were taking clozapine to have weekly or biweekly clinic visits to receive their medication, which would be likely to result in a higher adherence rate for these patients. At the time of the study, the hospital required that trials of at least two typical antipsychotics must have failed before a trial with an atypical agent could be initiated. Only patients entering the VA San Diego Healthcare System from an outside facility with a documented need for continued therapy with an atypical agent could continue the medication without fulfilling this criterion. In our study group, the large majority of patients who received an atypical agent had previous trials of two or more typical agents that had failed. Only 8.2% of the patients (N=14 of 171) had transferred to the VA San Diego Healthcare System while receiving an atypical antipsychotic.

Within the 3-month time period, we labeled the most recent prescription fill as the "index" fill. This initial group was trimmed to 629 prescriptions for 629 unique patients by eliminating duplicate prescriptions for individual patients and by excluding patients who received any combination of a typical plus an atypical antipsychotic medication. To minimize the heterogeneity of the patient pool, we further restricted the study group by applying the following inclusion criteria: 1) a chart diagnosis of DSM-IV schizophrenia, schizoaffective disorder, mood disorder with psychotic features, or psychosis not otherwise specified and 2) a minimum of two prescription fills within the 6 months before the index fill, accounting for at least a 60-day supply of antipsychotic medication. The diagnostic categories were chosen to represent patients who received continuing antipsychotic therapy for psychotic disorders. Patients were excluded if they received antipsychotics for nonpsychotic mood disorders or behavioral disturbances secondary to dementia or other general medical conditions. In addition, patients were excluded if their charts suggested that they were receiving even part of their medical care outside of the VA system. These selection criteria resulted in a study group of 288 patients, including 57 who were receiving haloperidol, 60 receiving perphenazine, 80 given risperidone, 63 given olanzapine, and 28 given quetiapine.

Data Collection

Demographic and relevant clinical information, including data on age, gender, ethnicity, psychiatric diagnosis, and psychotropic medications received, was collected. Data on prescriptions for antidepressants, mood stabilizers, and sedative-hypnotics were recorded for analysis of the use of adjuvant psychotropic medication. Data on prescriptions for medications to treat extrapyramidal symptoms were also recorded.

Adherence to prescribed regimens was determined by examining computerized medication fill records for a 12-month period. Refill assessments were carried out for up to 12 months beginning with the "baseline study fill" (i.e., the medication fill that occurred up to 6 months before the index fill). Adherence was computed with two previously described methods: 1) the cumulative mean gap ratio and 2) the compliant fill rate (19, 32).

The cumulative mean gap ratio for a specified time period was calculated by dividing the number of days that medication was unavailable for consumption (due to a delayed refill) by the total number of days during the time period (19, 32). The compliant fill rate for a specified time period represents the proportion of total fills that are adherent, i.e., filled at the appropriate time interval. Adherence was assessed by comparing the number of days when the antipsychotic was available for consumption (i.e., day supply) to the number of calendar days between fills. Fills obtained within a period equivalent to 80%–120% of the period covered by the previous prescription were considered adherent (19, 32). A compliant fill rate was determined for each patient on the basis of the assessable fills during the 12 months of observation. However, if a change in therapy, e.g., a dose change or a change to a new medication, required a new prescription to be filled before the end of the period covered by a previous prescription, the premature fill was deemed adherent. Thus, while the compliant fill rate was based on a series of dichotomous assessments of adherence, the cumulative mean gap ratio provided a continuous assessment that detected the magnitude of gaps in therapy and constituted a more clinically meaningful measure. The following are the formulae used in our refill record calculations:

Cumulative Mean Gap Ratio=([Total Number of Days in the Study Period – Total Number of Days that Medication was Available]/Total Number of Days in the Study Period) × 100

Compliant Fill Rate=(Number of Adherent Fills/Total Number of Fills) × 100

Data Analysis

Independent samples t tests were used to compare the mean age, number of adjuvant psychotropic medications, compliant fill rate, and cumulative mean gap ratio for patients receiving typical antipsychotics (haloperidol and perphenazine) versus those receiving atypical agents (risperidone, olanzapine, and quetiapine). If significant differences between the typical and atypical antipsychotic groups were found, one-way analysis of variance with Scheffé's post hoc tests was used to compare data for specific typical and atypical medications. Chi-square analysis was used to compare data on diagnoses, ethnicity, and gender between groups. Yates's corrections were employed for all two-by-two chi-square tests. To address our secondary study questions about the association of patient-related factors with nonadherence, Pearson product-moment correlation was used to examine the relationship between cumulative mean gap ratio data and age, daily dose of medication, and number of adjuvant psychotropic medications. Student's t tests were performed to evaluate the associa-

TABLE 1. Demographic and Clinical Characteristics of Outpatients Filling Prescriptions for Typical or Atypical Antipsychotic Medications Within a 12-Month Period in a Veterans Affairs (VA) Health Care System[a]

Characteristic	Patients Receiving Typical Antipsychotics (N=117)[b]		Patients Receiving Atypical Antipsychotics (N=171)[c]		Analysis		
	N	%	N	%	χ^2	df	p
Gender (male)	103	88.0	156	91.2	0.47	1	0.49
Ethnicity[d]					1.31	3	0.86
Caucasian	69	61.6	109	67.3			
African American	23	20.5	30	18.5			
Hispanic	12	10.7	15	9.3			
Other	8	7.1	8	4.9			
Diagnosis					1.96	3	0.58
Schizophrenia	62	53.0	91	53.2			
Schizoaffective disorder	26	22.2	45	26.3			
Mood disorder with psychotic features	14	12.0	21	12.3			
Psychosis not otherwise specified	15	12.8	14	8.2			
	Mean	SD	Mean	SD	t	df	p
Age (years)	49.2	11.0	50.1	11.9	0.60	286	0.55
Number of adjuvant psychotropic medications	1.5	1.0	1.4	1.0	-1.52	268	0.13

[a] Subjects were required to have a minimum of two prescription fills within the first 6 months of the 12-month study period, accounting for at least a 60-day supply of antipsychotic medication. Patients were excluded if they received clozapine, received a combination of typical and atypical antipsychotics, received an antipsychotic for a nonpsychotic mood disorder or for a behavioral disturbance secondary to dementia or other medical conditions, or had chart notes suggesting that some care was received outside the VA system.
[b] Typical antipsychotics comprised haloperidol and perphenazine.
[c] Atypical antipsychotics comprised risperidone, olanzapine, and quetiapine. Except for 14 patients who entered the VA system in continuing therapy with an atypical agent, patients were prescribed an atypical antipsychotic only after failed trials of at least two typical antipsychotics.
[d] N=112 for patients receiving typical antipsychotics; N=162 for patients receiving atypical antipsychotics.

tion of cumulative mean gap ratio data with ethnicity and diagnosis. Adherence rates were calculated at 6 and 12 months, but only 12-month adherence rates were used to answer our secondary study questions. All the statistical tests were two-tailed, and significance was defined as alpha=0.05.

Results

Patient and Treatment Characteristics

Comparisons among patients who received specific antipsychotic medications and between patients who received typical versus atypical medications revealed no significant group differences in gender, age, ethnicity, psychiatric diagnosis, and number of adjuvant psychotropic medications (Table 1). In the study group, 22.2% of the patients who received atypical antipsychotics (N=38) and 11.1% of patients who received typical agents (N=13) had at least one psychiatric hospitalization during the 1-year study period (χ^2=5.07, df=1, p=0.03). The mean total number of days of antipsychotic use analyzed per patient was 304.9 (SD=80.6). No differences existed among patients receiving specific medications in the mean total number of days of antipsychotic therapy analyzed (F=1.51, df=4, 283, p=0.20). The median total daily doses of antipsychotic medications, calculated by using each patient's highest prescribed dose for the 12-month period, were 8 mg of haloperidol (range=1–40), 12 mg of perphenazine (range=2–48), 4 mg of risperidone (range=0.5–12), 12.5 mg of olanzapine (range=5–30), and 400 mg of quetiapine (range=50–600).

Adherence Rates

On the basis of the cumulative mean gap ratio method, the patients with prescriptions for atypical antipsychotics had significantly smaller gaps in therapy compared to those with prescriptions for typical antipsychotics at 6 months (cumulative mean gap ratio=12.2%, SD=19.4%, compared with 22.9%, SD=25.9%) (t=–3.81, df=202, p= 0.001, unequal variances) and at 12 months (cumulative mean gap ratio=14.1%, SD=18.4%, compared with 23.2%, SD=25.0%) (t=–3.34, df=200, p=0.001, unequal variances). Thus, patients receiving typical agents were without medication for approximately 7 days per month, while those receiving atypical antipsychotics were without medication for approximately 4 days per month. Significant differences in the cumulative mean gap ratio existed at 6 months among some of the five medications included in the analysis (F=4.63, df=4, 283, p=0.001). Olanzapine had a significantly lower gap ratio (mean=10.3%, SD=19.8%) compared to haloperidol (mean=25.5%, SD=29.0%) (p=0.008, Scheffé). No other significant differences in cumulative mean gap ratio existed between individual antipsychotics at 6 or 12 months. Risperidone had a lower gap ratio (mean= 13.9%, SD=21.0%) than haloperidol at 6 months (mean= 25.5%, SD=29.0%), but the difference was not significant (p=0.06, Scheffé).

On the basis of the compliant fill rate method, the patients with prescriptions for atypical antipsychotics had a significantly higher adherence rate at 6 months (mean= 57.4%, SD=33.4%) than the patients with prescriptions for typical agents (mean=49.9%, SD=33.3%) (t=1.97, df=286,

FIGURE 1. Medication Adherence Rates at 12-Month Follow-Up for Outpatients Filling Prescriptions for Typical and Atypical Antipsychotic Medications in a Veterans Affairs (VA) Health Care System

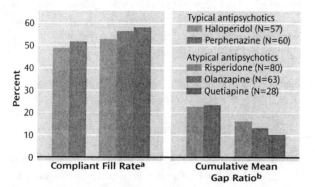

a Percentage of total medication fills that occurred at time-appropriate intervals. No significant difference between patients prescribed typical and atypical antipsychotics (F=0.84, df=4, 283, p=0.50).
b Percentage of total study days during which medication was unavailable because of a delayed refill. Significant difference between patients prescribed typical and atypical antipsychotics (F=3.61, df=4, 283, p=0.007). No significant differences between individual antipsychotics (p=0.12–1.00, Scheffé).

p=0.05). At 12 months, patients receiving atypical agents had a higher adherence rate (mean=54.9%, SD=26.0%) than those receiving typical agents (mean=50.1%, SD=30.6%), but the difference was not significant (t=1.59, df=286, p=0.11). No significant differences among individual antipsychotics in compliant fill rates were found at 6 or 12 months (Figure 1).

We also examined the proportions of patients in each group who had been receiving maintenance therapy (i.e., had received the same antipsychotic for at least 180 days before the study period). The typical antipsychotic group had more patients receiving maintenance therapy (80.4%, N=90 of 112) than the atypical antipsychotic group (69.0%, N=100 of 145); the difference approached but did not reach statistical significance (χ^2=3.68, df=1, p=0.055). When the data for patients who did not receive maintenance therapy were excluded, the 12-month adherence rates for typical and atypical agents (cumulative mean gap ratio: 21.0%, SD=23.9%, and 12.8%, SD=16.5%, respectively [t=–2.84, df=158, p=0.005, unequal variances]; compliant fill rate: 51.9%, SD=30.0%, and 55.8%, SD=25.2%, respectively [t=1.01, df=182, p=0.31, unequal variances]) were similar to the rates for the entire study group.

Factors Associated with Adherence

Age, total daily dose of antipsychotic, and number of adjuvant psychotropic medications were not significantly correlated with the cumulative mean gap ratio (r<±0.20, df=26–286, p=0.12–0.95 for all correlations). No differences in adherence were found between Caucasians and non-Caucasians (cumulative mean gap ratio=16.1%, SD=20.1%, versus 21.1%, SD=24.7%) (t=–1.71, df=161, p=0.09, unequal variances), between patients with psychosis with a mood component and those with schizophrenia or psychosis not otherwise specified (cumulative mean gap ratio=18.6%, SD=23.9%, versus 17.4%, SD=20.5%) (t=–0.44, df=286, p=0.66), or between patients who received anticholinergic medication for treatment of extrapyramidal symptoms at any time during the 12-month period of refill assessment and those who did not receive anticholinergics (cumulative mean gap ratio=19.5%, SD=22.2%, versus 16.9%, SD=21.5%) (t=0.94, df=286, p=0.35).

Discussion

We confirmed our hypothesis that patients with prescriptions for atypical antipsychotics would have a higher rate of adherence to their medication regimen than those receiving conventional agents. However, at 12 months, statistically significant differences in adherence between those groups were seen in only the cumulative mean gap ratio, and the significant differences in adherence for individual medications that were seen at 6 months (greater adherence for olanzapine than for haloperidol) were not present at 12 months. These results demonstrate the importance of longer adherence assessments. Although adherence rates were numerically highest for quetiapine, no definitive conclusions can be drawn because the number of subjects who received quetiapine was small. The refill rates observed in this study highlight the pervasive, problematic degree of antipsychotic nonadherence in patients with psychotic disorders, including those who have prescriptions for atypical agents.

As we postulated, we found no significant associations between nonadherence and age, gender, ethnicity, or diagnosis. We did not, however, confirm our hypothesis that patients taking higher total daily doses of antipsychotic or more adjuvant psychotropic medications would have lower rates of adherence.

Our findings are both similar to and different from those of Rosenheck et al. (14), Olfson et al. (16), and Cabeza et al. (17). The results for comparisons between adherence with typical antipsychotics and adherence with atypical antipsychotics in the previous studies ranged from no significant difference (17), to differences that were nonsignificant but approached significance (16), to a significant difference in one of the two measures of adherence (14). These discrepancies may be accounted for by differences in methods, including differences in adherence measures, definitions of adherence, specific medications included in the study, frequency of assessments, and treatment settings. The strengths of the present study include the length of assessment period (12 months) and the use of objective definitions of adherence based on refill records (the cumulative mean gap ratio and the compliant fill rate). Obtaining pharmacy refill information was an unobtrusive method of data collection, allowing a naturalistic estimate of adherence (18). In addition, the very low cost to patients of antipsychotic medications in the VA system likely minimized any effect of financial burden on refill rates, and the exclusion of patients who were likely to have received medical care outside the VA San Diego Healthcare System increased the likelihood that the pharmacy records were complete. Finally, it is noteworthy that the adherence rate among the patients with prescriptions for typical neuroleptics (compliant fill rate=50%) was very similar to the rates reported in previous reviews (1, 2).

The higher rates of adherence calculated for patients with prescriptions for atypical antipsychotics become more clinically meaningful when one considers the antipsychotic prescriptions policy in effect at the VA San Diego Healthcare System at the time this study was completed. Previous studies have reported that patients with more severe psychotic symptoms have higher rates of antipsychotic nonadherence (22, 30, 33). In our study group, significantly more patients who received atypical agents had one or more psychiatric hospitalizations, compared to the patients who received typical antipsychotics. The VA policy might have contributed to an underestimation of the effect of atypical antipsychotics on adherence because more patients with treatment-refractory illness and more treatment-intolerant patients may have had prescriptions for atypical agents. However, this explanation is speculative, and a plausible argument could be made to support the idea that the group receiving atypical antipsychotics included adherent patients because of factors that were unaccounted for, such as living situation and medication supervision status. Because we did not have data on psychopathology or side effects, we were unable to evaluate whether and to what degree these factors influenced rates of adherence. Additional studies will be needed to evaluate differences among the individual atypical antipsychotics in risk factors for nonadherence. Future studies should examine the clinical significance of the modest differences observed in adherence among typical and atypical antipsychotics.

We should point out several limitations of this study. For example, refill records do not directly measure medication intake. At the present time, however, no gold standard exists for the measurement of medication adherence (18). Moreover, all other methods of adherence assessment have specific limitations. Patient self-reports are by their very nature subjective. The accuracy of both pill counts and serum drug levels, measures that may seem less subject to bias, can also be influenced by patients' behavior, either intentionally or unintentionally. Newly developed performance-based tests to evaluate medication management skills, such as the Medication Management Ability Assessment, are promising, but do not measure adherence directly (34). Rates of adherence based on pharmacy refill records have been reported to correlate with other adherence behaviors (e.g., appointment keeping), serum drug levels, and drug effects such as blood pressure control (35–38). In addition, although refill records provided

an indirect measure of adherence, they enabled us to calculate gaps in therapy when patients were late in refilling their supply of medication, which demonstrated widespread underuse of medications. Finally, we attempted to improve the comprehensiveness of pharmacy refill records by excluding patients who received medical care outside of the VA system.

Another limitation was the nonrandomized nature of this study. Nonetheless, to reduce selection bias, we included all patients who met the selection criteria for this naturalistic study. Because the study used a retrospective design, we were unable to analyze the relationship of adherence to antipsychotic efficacy or side effects. In addition, although we did not find significant relationships between nonadherence and age, gender, ethnicity, or number of adjuvant psychotropic medications, we were unable to measure other potentially relevant patient-related factors, such as insight, medication supervision status, or substance abuse. Finally, our results may not generalize to non-VA patients.

In summary, on the basis of pharmacy refill records, VA outpatients with prescriptions for atypical antipsychotics had greater medication adherence, compared to patients with prescriptions for typical agents. Nonadherence was considerable, however, even among the patients receiving atypical antipsychotics. Future research in this area should focus on developing effective interventions to improve medication adherence in patients receiving antipsychotic medications.

Presented in part at the 14th annual meeting of the American Association for Geriatric Psychiatry, San Francisco, Feb. 24–26, 2001. Received Feb. 16, 2001; revision received June 18, 2001; accepted July 31, 2001. From the Department of Psychiatry, University of California, San Diego; and the Pharmacy and Psychiatry Services, Veterans Affairs (VA) San Diego Healthcare System. Address reprint requests to Dr. Dolder, Geriatric Psychiatry Intervention Research Center (116A-1), VA San Diego Healthcare System, 3350 La Jolla Village Dr., San Diego, CA 92161; cdolder@ucsd.edu (e-mail).

Supported in part by the Department of Veterans Affairs and NIMH grants MH-19934, MH-49671, MH-43693, and MH-59101. Dr. Lacro and Dr. Jeste have been consultants to and have received grant support for other projects from AstraZeneca, Bristol-Meyers-Squibb, Eli Lilly, Janssen, and Pfizer; however, no pharmaceutical industry support was received for this study.

The authors thank Shahrokh Gholshan, Ph.D., for assistance with the statistical analysis.

References

1. Fenton WS, Blyler CR, Heinssen RK: Determinants of medication compliance in schizophrenia: empirical and clinical findings. Schizophr Bull 1997; 23:637–651
2. Young JL, Zonana HV, Shepler L: Medication noncompliance in schizophrenia: codification and update. Bull Am Acad Psychiatry Law 1986; 14:105–122
3. Weiden PJ, Olfson M: Cost of relapse in schizophrenia. Schizophr Bull 1995; 21:419–429
4. Terkelsen KG, Menikoff A: Measuring the costs of schizophrenia: implications for the post-institutional era in the US. Pharmacoeconomics 1995; 8:199–222

5. Wyatt RJ, Henter I, Leary MC, Taylor E: An economic evaluation of schizophrenia: 1991. Soc Psychiatry Psychiatr Epidemiol 1995; 30:196–205

6. Rice DP: The economic impact of schizophrenia. J Clin Psychiatry 1999; 60:4–6

7. Kane J, Honigfeld G, Singer J, Meltzer H (Clozaril Collaborative Study Group): Clozapine for the treatment-resistant schizophrenic: a double-blind comparison with chlorpromazine. Arch Gen Psychiatry 1988; 45:789–796

8. Marder SR, Meibach RC: Risperidone in the treatment of schizophrenia. Am J Psychiatry 1994; 151:825–835

9. Tollefson GD, Beasley CM Jr, Tamura RN, Tran PV, Potvin JH: Blind, controlled, long-term study of the comparative incidence of treatment-emergent tardive dyskinesia with olanzapine or haloperidol. Am J Psychiatry 1997; 154:1248–1254

10. Small JG, Hirsch SR, Arvanitis LA, Miller BG, Link CG: Quetiapine in patients with schizophrenia: a high- and low-dose double-blind comparison with placebo. Arch Gen Psychiatry 1997; 54:549–557

11. Jeste DV, Lacro JP, Bailey A, Rockwell E, Harris J, Caligiuri MP: Lower incidence of tardive dyskinesia with risperidone compared with haloperidol in older patients. J Am Geriatr Soc 1999; 47:716–719

12. Jeste DV, Rockwell E, Harris MJ, Lohr JB, Lacro J: Conventional vs newer antipsychotics in elderly patients. Am J Geriatr Psychiatry 1999; 7:70–76

13. Marder SR: Facilitating compliance with antipsychotic medication. J Clin Psychiatry 1998; 59:21–25

14. Rosenheck R, Chang S, Choe Y, Cramer J, Xu W, Thomas J, Henderson W, Charney D: Medication continuation and compliance: a comparison of patients treated with clozapine and haloperidol. J Clin Psychiatry 2000; 61:382–386

15. Arana GW: An overview of side effects caused by typical antipsychotics. J Clin Psychiatry 2000; 61:5–11

16. Olfson M, Mechanic D, Hansell S, Boyer CA, Walkup J, Weiden PJ: Predicting medication noncompliance after hospital discharge among patients with schizophrenia. Psychiatr Serv 2000; 51:216–222

17. Cabeza IG, Amador MS, Lopez CA, Chavez MG: Subjective response to antipsychotics in schizophrenic patients: clinical implications and related factors. Schizophr Res 2000; 41:349–355

18. Vitolins MZ, Rand CS, Rapp SR, Ribisl PM, Sevick MA: Measuring adherence to behavioral and medical interventions. Control Clin Trials 2000; 21:188S–194S

19. Steiner JF, Prochazka AV: The assessment of refill compliance using pharmacy records: methods validity, and applications. J Clin Epidemiol 1997; 50:105–116

20. Buchanan RW, Breier A, Kirkpatrick B, Ball P, Carpenter WT Jr: Positive and negative symptom response to clozapine in schizophrenic patients with and without the deficit syndrome. Am J Psychiatry 1998; 155:751–760

21. Buchanan A: A two-year prospective study of treatment compliance in patients with schizophrenia. Psychol Med 1992; 22:787–797

22. Dixon L, Weiden P, Torres M, Lehman A: Assertive community treatment and medication compliance in the homeless mentally ill. Am J Psychiatry 1997; 154:1302–1304

23. Heyscue BE, Levin GM, Merrick JP: Compliance with depot antipsychotic medication by patients attending outpatient clinics. Psychiatr Serv 1998; 49:1232–1234

24. Nageotte C, Sullivan G, Duan N, Camp PL: Medication compliance among the seriously mentally ill in a public mental health system. Soc Psychiatry Psychiatr Epidemiol 1997; 32:49–56

25. Owen RR, Fischer EP, Booth BM, Cuffel BJ: Medication noncompliance and substance abuse among patients with schizophrenia. Psychiatr Serv 1996; 47:853–858

26. Pan PC, Tantam D: Clinical characteristics, health beliefs and compliance with maintenance treatment: a comparison between regular and irregular attenders at a depot clinic. Acta Psychiatr Scand 1989; 79:564–570

27. Agarwal MR, Sharma VK, Kumar K, Lowe D: Non-compliance with treatment in patients suffering from schizophrenia: a study to evaluate possible contributing factors. Int J Soc Psychiatry 1998; 44:92–106

28. Razali MS, Yahya H: Compliance with treatment in schizophrenia: a drug intervention program in a developing country. Acta Psychiatr Scand 1995; 91:331–335

29. Garavan J, Browne S, Gervin M, Lane A, Larkin C, O'Callaghan E: Compliance with neuroleptic medication in outpatients with schizophrenia; relationship to subjective response to neuroleptics; attitudes to medication and insight. Compr Psychiatry 1998; 39:215–219

30. Duncan JC, Rogers R: Medication compliance in patients with chronic schizophrenia: implications for the community management of mentally disordered offenders. J Forensic Sci 1998; 43:1133–1137

31. Smith TE, Hull JW, Goodman M, Hedayat-Harris A, Willson DF, Israel LM, Munich RL: The relative influences of symptoms, insight, and neurocognition on social adjustment in schizophrenia and schizoaffective disorder. J Nerv Ment Dis 1999; 187:102–108

32. Hamilton RA, Briceland LL: Use of prescription-refill records to assess patient compliance. Am J Hosp Pharm 1992; 49:1691–1696

33. Corriss DJ, Smith TE, Hull JW, Lim RW, Pratt SI, Romanelli S: Interactive risk factors for treatment adherence in a chronic psychotic disorders population. Psychiatry Res 1999; 89:269–274

34. Patterson TL, Lacro J, McKibbin CL, Davidson K, Jeste DV: Direct observation of medications management (MMAA): results from a new performance-based test in older outpatients with schizophrenia. J Clin Psychopharmacol (in press)

35. Deyo RA, Inui TS, Sullivan B: Noncompliance with arthritis drugs: magnitude, correlates, and clinical implications. J Rheumatol 1981; 8:931–936

36. Peterson GM, McLean S, Millingen KS: Determinants of patient compliance with anticonvulsant therapy. Epilepsia 1982; 23:607–613

37. Steiner JF, Koepsell TD, Fihn SD, Inui TS: A general method of compliance assessment using centralized pharmacy records. Med Care 1988; 26:814–823

38. Steiner JF, Fihn SD, Blair B, Inui TS: Appropriate reductions in compliance among well-controlled hypertensive patients. J Clin Epidemiol 1991; 44:1361–1371

"Does the Payment of Incentives Create Expectation Effects," Eleanor Singer, John Van Hoewyk, Mary P. Maher, *Public Opinion Quarterly*, Vol. 62, Issue 2, pp. 152-164.

Background

Some surveys are conducted over a period of many years. It is important to have a group of subjects who participate in subsequent surveys, so that you can tell how individuals have changed over time. At the same time, many people are reluctant to participate in surveys. To get people to participate, they are offered a relatively small amount of money. The concern is that these people will then need to be paid to participate in subsequent surveys.

Questions

1. What is the objective of this study? Describe the data they used. What was their sampling scheme?
2. Table 1 on p. 157 describes some of the data. What did the authors learn from this table? What test did they do on the data? Describe an alternative test. Why can't this alternative be applied to the responses to the second question? At the bottom of the table, footnotes identify three responses as being significant. What are the actual p-values for these tests?
3. What did the authors learn from Table 2? Do you agree with their conclusions? On
4. p. 159, the authors compare results for people who did not get paid in either year. They state that "p=.18." Explain what this means and how it fits in with the authors' conclusions.
5. Discuss what the authors learned from Table 3. What test(s) did they do? On the bottom of p. 159, the authors state that there is a significant difference between the first two columns of Table 3. However, there is not a significant difference between the first column and the third column. But the participation rate for the second and third columns is nearly the same (81% and 79.5%). Why is one comparison significant while the other is not? Also, what is the p-value for comparing the first and third columns?
6. Table 4 concerns "Bivariate Regressions." What are the response and predictor variables for each regression? Most of the coefficients listed are negative. Why is this? The first coefficient listed is minus .01 for Respondent should get paid in 1996. Interpret this value. The number of asterisks indicates the significance level of the regression. How would you compute these significance levels? Most of the SE for coefficients are 0.02 or so. The exception is Will do again in 1996, where SE is 0.21. What does this tell you? What about the fact that this coefficient is marked with three asterisks? Give a likely explanation for this SE.
7. At the top of p. 162, the authors discuss some further regression results. What role does this discussion play in the overall aims of the study? Do the p-values cited appear correct? How would you calculate them? The SE values for these coefficients are quite a bit higher than in Table 4. Why is this?

DOES THE PAYMENT OF INCENTIVES CREATE EXPECTATION EFFECTS?

ELEANOR SINGER
JOHN VAN HOEWYK
MARY P. MAHER

Abstract Increasing use of incentive payments to survey respondents raises the threat of several unintended consequences, among them the creation of expectations for future payments and the possibility of a deterioration in the quality of response. Such deterioration may come about either as a direct result of substituting external for internal motivation, or as a consequence of expectations for rewards that go unmet by the survey organization. The findings from the present study are somewhat reassuring with respect to both of these unintended outcomes. Although people who have received a monetary incentive in the past are significantly more likely than those who have not to endorse the statement that ''people should be paid for doing surveys like this,'' they are actually more likely to participate in a subsequent survey, in spite of receiving no further payments. And respondents who received an incentive 6 months earlier are no more likely than those who received no incentive to refuse to answer (or to answer ''don't know'' to) a series of 18 key questions on the survey. Furthermore, they are more likely than other respondents to express favorable attitudes toward the usefulness of ''surveys like this.'' The generality of these findings, however, needs much further testing.

There is some evidence that the difficulty of obtaining cooperation with sample households in the United States and other developed countries is growing over time (de Heer and Israëls 1992). Even though the overall response rates of surveys may be relatively constant (Smith 1995), the

ELEANOR SINGER is senior research scientist, JOHN VAN HOEWYK is senior research associate, and MARY P. MAHER is senior survey specialist, all at the Survey Research Center, Institute for Social Research, University of Michigan. An earlier version of this article was presented at the Ninth Nonresponse Workshop, Mannheim, Germany, September 24, 1997. The authors would like to thank their colleague, Mick P. Couper, for his helpful comments, and the Survey Research Center for financial support of this research.

Public Opinion Quarterly Volume 62:152–164 © 1998 by the American Association for Public Opinion Research
All rights reserved. 0033-362X/98/6202-0002$02.50

component of nonresponse due to refusals appears to be increasing unless efforts by survey organizations are substantially increased (Groves and Couper 1996).

In an effort to counter the increasing problem of noncooperation, survey organizations are offering incentives to respondents with increasing frequency, some at the outset of the survey, as has traditionally been done in mail surveys, and some only after the person has refused, in an attempt to convert the refusal. In the case of mail surveys, the payment of incentives is one of two design factors that consistently and substantially increase the response rate, the other being the number of contacts (Heberlein and Baumgartner 1978; Yu and Cooper 1983). A meta-analysis by Church (1993) identified those characteristics of incentives in mail surveys that are associated with greater effects on response rates: prepayment, cash, and larger (vs. smaller) payments. A subsequent examination of the use of incentives in telephone and face-to-face surveys (Singer et al., in press) demonstrated the utility of incentives in those surveys, as well. There appear to be no deleterious effects of incentives on the quality of survey responses, though further research is needed in this area.

Despite these encouraging findings, concerns persist about possible unintended consequences of the use of incentives. Three can be mentioned here.

One is a concern that the use of differential incentives to convert refusals will be perceived as unfair by respondents, and will adversely affect their attitudes toward surveys and their willingness to cooperate. Research on this question has resulted in mixed findings. A laboratory study using videotaped vignettes of doorstep survey requests (Groves et al. 1997) found, as predicted, that disclosure of refusal conversion payments to some respondents but not others resulted in significantly lower expressed willingness, on the part of subjects viewing the tape, to participate in the survey. However, a replication of this portion of the study in a field setting failed to reproduce this key finding, even though approximately three-quarters of subjects in the lab and three-quarters of those in the field experiment judged the practice of refusal conversion payments to be unfair.[1] Although further research would be useful in clarifying the reasons for the differences between the two studies, it would appear that equity issues are not highly relevant to the decision whether or not to participate in a survey.

1. There were differences between these studies in the dependent variable. In the lab, subjects were asked whether they would be willing to participate in the survey that they had just seen described in the video vignette. In the field study, subjects were asked, at the conclusion of an actual face-to-face interview, whether they would be willing to participate in a future survey by the same survey organization. These differences may account for the difference in results between the two studies. For a further discussion, see Singer, Groves, and Corning (1997).

A second issue that has aroused some concern among survey researchers is whether the offer of an incentive is likely to replace intrinsic motivation to participate with extrinsic motivation, with a resulting decline in the quality of response. Research with children (Deci 1971; Lepper, Greene, and Nisbett 1973) has demonstrated such a motivational shift in regard to play activities, for example. In a study of how framing an incentive affects response, Singer et al. (1997) found suggestive evidence that students who respond to a survey request following receipt of a small gift perceive themselves as having responded primarily because of interest, whereas those who responded following receipt of a check for $10 perceive themselves as having responded primarily because of the incentive. (An analogous finding is reported by Lengacher et al. [1995], who found that the usual measure of enjoyment of the interview is less predictive of wave 2 participation among respondents who had received a substantial refusal conversion payment in wave 1 than among those who had received no such incentive.) However, since relatively few people in the general population can be counted on to have an intrinsic interest in surveys, the appeal to extrinsic motivation may be inevitable in this case. As noted above, studies that have addressed the response quality issue have so far found no differences between those who do and those who do not receive an incentive, but very few studies have systematically investigated this issue, and they have investigated only limited aspects of quality.

Still a third concern is that payment of incentives, especially at the outset of a study, may lead to expectations for such incentives in future surveys. The prediction that offering an incentive to respondents is likely to lead to increased expectations for such incentives in the future can be derived from theories that predict the development of norms (expectations) from perceptions of the existing state of affairs. Such theories include expectation-state theory (Berger et al. 1972) and several versions of exchange theory (Cook 1975; Homans 1974) as well as Davis's functional theory (1948). These theories are largely silent on the consequences of the development of norms for the quality of behavior, although they all suggest that the violation of expectations is likely to evoke a negative reaction (but see Cook [1975]).

In an investigation of the effects of differential incentives in one wave on participation in a later wave of a longitudinal survey, Lengacher and her colleagues (1995) found no effect of a large refusal conversion payment on subsequent participation, compared to other wave 1 reluctant respondents. In this article, we provide further evidence bearing on two hypothesized unintended consequences of the payment of incentives—namely, increased expectations of being rewarded in the future as a result of having been rewarded in the past, and declines in the quality of response.

Methods

Because of concerns about declining cooperation with survey requests, the Survey Research Center at the University of Michigan decided in the fall of 1995 to begin monitoring the changing climate for survey research in the United States by adding five evaluative questions at the end of the Survey of Consumer Attitudes (SCA), a national telephone survey administered monthly to a sample of roughly 500 respondents. Of these, 300 are newly selected random-digit-dialed households, and the remaining 200 are reinterviews of respondents first interviewed 6 months earlier. Because of concerns about their possible biasing effects, the five evaluative questions were asked of only the reinterviewed portion of the sample. The questions were added to the survey in January and February of 1996 and repeated in February and March of 1997 in order to measure changes in the climate for survey research over the approximately 12-month period. In principle, one could use changes in the responses to these questions as leading indicators of changes in the climate for survey research, and take proactive steps to counteract such changes.

The five monitoring questions, which were systematically rotated during their administration, are as follows:

1. If you had it to do over again, would you have agreed to do the interview or would you have refused?

For each of the following, please tell me whether you agree strongly, agree somewhat, disagree somewhat, or disagree strongly:

2. Surveys like this one provide useful information for decision makers.
3. Surveys like this one are a waste of people's time.
4. People should get paid for doing surveys like this.
4a. How much should they get paid?
5. Everyone has a responsibility for answering surveys like this. (Do you agree strongly, agree somewhat, disagree somewhat, or disagree strongly?)

The initial response rate of those reinterviewed in January and February of 1996 was 67 percent 6 months earlier; their reinterview rate averaged 77.6 percent. Thus, the effective response rate of the 1996 sample is 52 percent. For the sample reinterviewed in February and March of 1997, the initial response rate averaged 65.3 percent and the reinterview rate, 76.8 percent; thus, the effective response rate for the 1997 sample was a slightly lower 50.2 percent. (The response rate excludes only nonsample cases from the denominator, and is thus a fairly conservative estimate. Noninterviews for reasons of illness or language, e.g., are retained in the denominator.) The monitoring questions are, thus, asked primarily of cooperative respondents and of those who are easier to reach at home, and the comparisons discussed in this article might be somewhat different if it had been possible to include nonrespondents.

Of particular importance for the present study, approximately half the respondents to the March 1997 survey had been promised a $5 incentive in return for their participation 6 months earlier, as part of a randomized experiment. In addition, a much smaller number of respondents in three of the four months had received refusal conversion payments of $20–$25 6 months earlier.[2] Thus, we are able to evaluate the effect of incentives on subsequent attitudes and behavior, and to do so in the context of what was essentially a randomized experiment for the large majority of respondents.[3]

Results

CHANGES IN EXPECTATIONS ABOUT PAYMENT FOR SURVEY PARTICIPATION

As already noted, the reason for asking these questions at two points in time was to capture changes in the climate for surveys. Responses to the five questions (and one subquestion) in 1996 and 1997 are shown in table 1. They represent two cross-sectional measurements of attitudes toward surveys rather than answers by the same respondents at two different times.

Table 1 indicates that on three of the questions, no significant changes took place from one year to the next. Three others, however, show a significant change. Significantly more respondents (45.7 percent in 1997, compared to 29.7 percent in 1996) said that respondents should be paid for doing a survey like this, and the amount they stipulated showed a significant increase as well. In addition, a significantly higher proportion of respondents said, in 1997, that everyone has a responsibility for doing surveys like this.

We had anticipated changes in answers to the question about payment for two reasons, both derived from the theoretical argument above. First, the practice of paying incentives to respondents in telephone and face-to-face surveys appears to be increasing. To the extent that awareness of this is diffusing throughout the population, a generalized expectation for payment may be developing. Second, as already noted, a large number of respondents to the March 1997 survey had themselves received a $5 initial incentive payment 6 months earlier, and a smaller number of respondents in three of the four months had received a refusal conversion

2. Since no one had been offered an initial incentive before 1996, we cannot estimate the effect of such payments on changes in expectations.
3. Interviewers were blind to the fact that respondents had been offered an initial incentive 6 months earlier; however, they did know whether the respondent had received a refusal conversion payment.

Table 1. Responses to Five Evaluative Questions
(by Year)

Question	1996 (%)	1997 (%)
Respondent should get paid:**		
Agree	29.7	45.7
Disagree	70.3	54.3
(N)	(411)	(396)
How much:*		
0–5	19.6	22.2
6–10	32.4	19.4
11–20	19.6	35.4
Over 20	28.4	22.9
(N)	(102)	(144)
Surveys are useful:		
Agree strongly	34.6	39.7
All other	65.4	60.3
(N)	(405)	(401)
Everyone's responsibility:*		
Yes	44.5	51.4
No	55.5	48.6
(N)	(409)	(403)
Will do again:		
Yes	76.9	76.1
No	23.1	23.9
(N)	(407)	(406)
Waste of time:		
Disagree strongly	28.6	30.4
All other	71.4	69.6
(N)	(402)	(395)

* $p < .05$.
** $p < .01$.

payment of $20–$25. These respondents might have developed an expectation for payment based on their personal experience.

In order to separate the effect of these two reasons—generalized expectations versus personal experience—we looked at the responses to the "people should get paid" question among those who had and those who had not been offered an incentive. The results are shown in table 2. In both years, those who had received an incentive were much more likely to say that people should be paid than those who had not; the differences are significant in both years, and are especially large for people who re-

Table 2. Responses to Evaluative Questions, by Year and Receipt of Incentive

	1996		1997		
	0	$20–$25	0	$5	$20–$25
Respondent should get paid (%)	26.0	91.3***	31.0	51.3	77.6***
Surveys are useful (%)	32.9	63.6***	35.2	48.7	40.6
Everyone's responsibility (%)	44.8	39.1	45.7	59.0	58.0*
Will do again (%)	77.6	65.2	74.5	83.3	76.7
Waste of time (disagree strongly) (%)	26.6	63.6***	30.0	38.5	26.5
(N)	(393)	(23)	(242)	(81)	(70)

*p < .10.
*** p < .01.

ceived the (larger) refusal conversion payments. Among people who did not receive any kind of incentive in either year, there is a slight increase from 1996 to 1997 in the percentage saying people should get paid for doing surveys like this, but this difference is not significant; $\chi^2 = 1.77$, $df = 1$, $p = .18$. Among people who did receive an incentive as a refusal conversion payment, the direction of change is actually reversed; but the difference between 1996 and 1997 is not statistically significant. Thus, table 2 demonstrates that the changed expectations apparent in table 1 are due almost entirely to the responses of those who had themselves received an incentive—in other words, to personal experience rather than diffuse social norms.

The question of interest for this article, however, is what interpretation should be placed on these responses. Should they be understood as reflecting changed expectations for the future, or rather as normative statements justifying past behavior? Dissonance theory (Festinger 1957) would lead us to expect that people who had themselves accepted payment 6 months earlier should be unlikely to challenge the legitimacy of this behavior; that is, they should be likely to agree with the statement that people should be paid.

We can test these alternative interpretations by examining the cooperation rate of people to the March 1997 survey. Among people recontacted in March 1997, 139 had received an incentive 6 months earlier and 98 had not; 28 received a refusal conversion payment in March. If the earlier payment of an incentive led to (unmet) expectations for payment in the future, we would expect cooperation rates (without an incentive) in March to be lower among those who had received an initial incentive the preceding September than among those who had not. However, among those who had received an initial incentive in September and who were contacted by interviewers, 81 percent were reinterviewed without an additional incentive in March; among those who had received a refusal conversion incentive in September, the cooperation rate in March without an additional incentive was 79.5 percent; and among those who had received no incentive in September, the cooperation rate without an additional incentive in March was 66.3 percent (see table 3). The difference between those receiving no incentive in September and those receiving \$5 is significant: $\chi^2 = 5.43$, $df = 1$, $p < .05$. Those who had received a \$5 incentive 6 months earlier were significantly more likely to cooperate in March than those who had received no incentive. The difference in March cooperation between the no-incentive group and the group receiving a refusal conversion payment in September is not significant. Thus, these data provide no evidence that responses to the question about whether or not respondents should get paid reflect expectations about future payment.

Three alternative interpretations of the results in table 3 might be entertained: (1) the responses to the payment question reflect justifications of

Table 3. March 1997 Cooperation by September 1996
Incentive Status

	September Incentive		
March Cooperation	0	$5	$20–$25
Interviewed, no incentive (%)	66.3	81.0	79.5
(*N*)	(98)	(100)	(39)

past behavior; (2) the responses reflect expectations about payment for participation in surveys by organizations other than the Survey Research Center; (3) the responses reflect expectations about payment by the Survey Research Center, but for a survey other than this one. We return to these speculations below.

Because the increase in the percentage saying people should be paid seems to conflict with the increased tendency, also documented in table 1, to say that everyone has a responsibility to participate in ''surveys like this,'' we cross-tabulated the responses to these two questions in both years. The association is significant in neither year. In both years, people who agree that respondents should be paid for doing a survey like this are neither more nor less likely than those who disagree to say that everyone has a responsibility for participating in such a survey.[4] Nor were there any significant associations between the judgment that respondents should be paid and responses to any of the other monitoring questions.

EXPECTATIONS OF PAYMENT, PAYMENT, AND DATA QUALITY

As noted earlier, the theory linking incentive payments and survey response rates suggests that researchers may pay a price in quality for the increased participation rate made possible by incentives. In this section, we first examine the relationship between expectations of payment and an indicator of the quality of response, and then we look at actual payment and the same indicator.

The indicator of response quality used in this article is an index of

4. The survey is introduced to respondents as a survey conducted by the University of Michigan ''about the economy.'' At their discretion, interviewers may add, ''It's part of the Index of Leading Economic Indicators for the Commerce Department—the Consumer Confidence Report.'' If respondents ask, they are told that the results are made available to policy makers in government and business and are cited in newspapers, magazines, and on television, that the survey has been an ongoing project at the university for nearly 50 years, and that it is used by government, business, and academic economists to forecast and understand the nation's economy.

Table 4. Bivariate Regressions of Nonresponse Index on Responses to Five Evaluative Questions (by Year)

	1996		1997	
Question	Coefficient	SE	Coefficient	SE
Respondent should get paid	−.01	(.02)	−.00	(.02)
Surveys are useful	−.07	(.02)***	−.05	(.02)**
Everyone's responsibility	−.04	(.02)*	−.02	(.02)
Will do again	−.10	(.21)***	−.04	(.02)*
Waste of time (disagree strongly)	−.07	(.03)***	−.05	(.03)*

*$p < .10$.
**$p < .05$.
***$p < .01$.

nonresponse to key questions on the questionnaire.[5] This indicator of response quality behaves in predictable ways in relation to four of the five monitoring questions (see table 4). Respondents who say they would not do the survey again, who do not consider surveys like this one useful, who consider them a waste of time, and who disagree that everyone has a responsibility to take part in surveys "like this" are significantly ($p <$ .10) more likely to answer "don't know" or to refuse to answer key substantive items on the questionnaire, although the regression predicting item nonresponse from answers to the question about responsibility is not significant in 1997. The fact that these items behave as predicted in relation to a behavioral indicator of quality makes the absence of such a relationship between expectations of payment and the index of nonresponse more striking. In these data, the response that people should be paid for survey participation appears to have no negative consequences for the quality of response.

PAYMENT OF INCENTIVES AND ATTITUDES TOWARD SURVEYS

We also examined the effect of the actual payment of incentives, whether offered at the outset of the study or as a refusal conversion payment, on the quality of responses, as measured by the nonresponse index. We found no effect of either refusal conversion payments or initial incentives on

5. The index of nonresponse is the percentage of "don't know" and "no answers" to 18 key questions in the Survey of Consumer Attitudes. The questions, whose tabulated responses appear in each SCA monthly report, indicate, among other things, respondents' assessment of their current and future family finances and income, the nation's business and employment conditions, and the government's role in affecting the country's economy.

the index. For 1996, when only refusal conversion payments were offered, $b = .90$, SE $= 1.29$, $p = .49$; for 1997 refusal conversion payments, $b = -.03$, SE $= .62$, $p = .96$; for initial incentives, $b = -.24$, SE $= .66$, $p = .71$.

As an additional indicator of response quality, we also examined the distribution of substantive responses to the 18 questions on four separate survey months in which incentive experiments had been carried out. Only four of the 72 comparisons showed a significant difference ($p < .10$), depending on whether or not the respondent had been offered an initial incentive—a number well within the limits of chance. Thus, there is no evidence that the offer of an initial incentive influences either item nonresponse or the distribution of substantive responses.

In both years, the payment of incentives affected responses to two of the five evaluative questions in addition to whether or not respondents should get paid (see table 2). In 1996, respondents who had 6 months earlier received refusal conversion payments were significantly more likely to say surveys are useful and to disagree that they are a waste of time. In 1997, respondents who had received any type of incentive 6 months earlier were significantly more likely to agree that surveys are useful and to say that everyone has a responsibility to take part in surveys like this. Thus, payment of incentives seems to lead to more favorable attitudes toward surveys, at least "surveys like this"—that is, the one for which the respondent has received payment. With one exception, there were no significant relationships between responses to the question about incentives and any of the demographic characteristics we examined. In 1996, but not 1997, nonwhites were significantly more likely than whites to say that respondents should be paid for doing a survey like this (data not shown).[6]

Discussion and Conclusion

Increasing use of incentive payments to survey respondents raises the threat of several unintended consequences, among them the creation of expectations for future payments and the possibility of a deterioration in the quality of response. Such deterioration may come about either as a direct result of substituting external for internal motivation, or as a consequence of expectations for rewards that go unmet by the survey organization.

The findings from the present study are somewhat reassuring with respect to both of these unintended outcomes. Although people who have

---

6. "Nonwhites" includes 155 blacks, 120 Hispanics, and 55 American Indians, Alaskan natives, Asians, or Pacific Islanders.

received a monetary incentive in the past are significantly more likely than those who have not to endorse the statement that "people should be paid for doing surveys like this," they are actually more likely to participate in a subsequent wave of the survey, even when they receive no further payments. Thus, it may be that respondents interpret the earlier payment as covering their current participation, as well. Respondents who received an incentive 6 months earlier are no more likely than those who received no incentive to refuse to answer (or to answer "don't know" to) a series of 18 key questions on the survey. Furthermore, they are more likely than other respondents to express favorable attitudes toward the usefulness of surveys "like this." We found no indication that the offer of an initial incentive affects the distribution of substantive responses.

The results of the present study are not, however, grounds for complacency. Payment of incentives is still a rather novel experience for respondents to telephone or personal interviews. Although few organizations would undertake a mail survey without enclosing some monetary or nonmonetary incentive with the advance letter or the questionnaire, interviewer-mediated surveys most commonly reserve incentives for refusal conversion efforts. Whether the absence of negative results observed in the present study will survive the more widespread use of incentives in such surveys remains very much an open question, one deserving continued research. At issue, as well, is whether the absence of negative effects extends to the surveys of other organizations or other surveys by the same organization, or whether this finding is limited to the second wave of a study for which the respondent has already received payment. Further research to answer these questions is currently under way.

References

Berger, J. M., M. Zelditch, Jr., B. Anderson, and B. Cohen. 1972. "Structural Aspects of Distributive Justice: A Status Value Formulation." In *Sociological Theories in Progress,* vol. 2, ed. Joseph Berger, Morris Zelditch, Jr., and Bo Anderson, pp. 119–46. Boston: Houghton Mifflin.

Church, A. H. 1993. "Estimating the Effect of Incentives on Mail Survey Response Rates: A Meta Analysis." *Public Opinion Quarterly* 57:62–79.

Cook, K. S. 1975. "Expectations, Evaluations, and Equity." *American Sociological Review* 40:372–88.

Davis, K. 1948. *Human Society.* New York: Macmillan.

Deci, E. L. 1971. "The Effects of Externally Mediated Rewards on Intrinsic Motivation." *Journal of Personality and Social Psychology* 18:105–15.

De Heer, W. F., and A. Z. Israëls. 1992. "Nonresponse Trends in Europe." Paper presented at the Joint Statistical Meetings of the American Statistical Association, Boston.

Groves, R. M., and M. P. Couper. 1996. "Household-Level Determinants of Survey Nonresponse." In *Advances in Survey Research,* ed. M. T. Braverman and J. K. Slater. New Directions for Evaluation, no. 70. San Francisco: Jossey-Bass.

Groves, R. M., E. Singer, A. Corning, and A. Bowers. 1997. "The Effects on Survey

Participation of Interview Length, Incentives, Differential Incentives, and Refusal Conversion: A Laboratory Approach." Paper presented at the annual meeting of the American Association for Public Opinion Research, Norfolk, VA.

Heberlein, T. A., and R. Baumgartner. 1978. " Factors Affecting Response Rates to Mailed Questionnaires: A Quantitative Analysis of the Published Literature." *American Sociological Review* 3:447–62.

Homans, G. C. (1961) 1974. *Social Behavior: Its Elementary Forms.* New York: Harcourt Brace Jovanovich.

Lengacher, J. E., C. M. Sullivan, M. P. Couper, and R. M. Groves. 1995. "Once Reluctant, Always Reluctant? Effects of Differential Incentives on Later Survey Participation in a Longitudinal Study." Paper presented at the annual meeting of the American Association for Public Opinion Research, Fort Lauderdale, FL, May.

Lepper, M. R., D. Greene, and R. E. Nisbett. 1973. "Undermining Children's Intrinsic Interest with Extrinsic Rewards." *Journal of Personality and Social Psychology* 28: 129–37.

Singer, E., N. Gebler, T. Raghunathan, J. Van Hoewyk, and K. McGonagle. 1999. "The Effects of Incentives on Response Rates in Personal and Telephone Surveys: Results from a Meta Analysis." *Journal of Official Statistics,* vol. 15 (in press).

Singer, E., N. Gebler, J. Van Hoewyk, and J. Brown. 1997. "Does $10 Equal $10? The Effect of Framing on the Impact of Incentives." Paper presented at the annual meeting of the American Association for Public Opinion Research, Norfolk, VA.

Singer, E., R. M. Groves, and A. Corning. 1997. "Incentives: The Effect of Perceived Equity on Willingness to Participate in a Future Survey." Survey Research Center, University of Michigan.

Smith, T. W. 1995. "Trends in Non-Response Rates." *International Journal of Public Opinion Research* 7:157–69.

Yu, J., and H. Cooper. 1983. "A Quantitative Review of Research Design Effects on Response Rates to Questionnaires." *Journal of Marketing Research* 20:36–44.

"Efficacy of St. John's Wort Extract WD5570 in Major Depression: A Double-Blind, Placebo-Controlled Trial," Y. Lecrubier, MD, G. Clerc, MD, R. Didi, MD, M. Kieser, PhD, *American Journal of Psychiatry*, Vol. 159, pp. 1361-1366.

Background

Depression is a mental illness that affects many people. There are medications that can be useful, but some people prefer not to take them, in part due to side effects. At the same time, there has been an increased interest in herbal treatments. St. John's Wort has often been cited as an antidepressant. This study wants to determine how well it work, or its "efficacy."

Questions

1. What was the objective of this study? What variables were measured to attain this objective? Identify all the variables in the study and classify as categorical or quantitative.

2. Table 1 gives some baseline data on the participants. No statistical analysis was given. How would you test if the gender breakdown of the two groups was different? How would you test Severity of Illness? How about age, Hamilton depression rating, Montgomery-Asberg rating and SCL-58 score?

3. Table 1 gives the range for the bottom four variables. How useful is this information? Can it tell you whether the data is approximately normal? What information would be more useful? How much do you need to know about the distribution of the data?

4. Why did the authors present Table 2? What were they trying to ascertain? What conclusion did they draw? Do you agree?

5. In Table 2, there are two footnotes about significant differences. Interpret the t-values. How can there be a significant difference overall and for score ≥ 22, but not for score < 22? The mean difference for the group under 22 was actually larger than the difference for the group over 22, but the first difference was not significant. How can that be? The footnote states that a two-sided test was done. Was this appropriate? If a one-sided test had been done, what would have changed, both in terms of the statistics and in terms of the conclusions?

6. On p. 1364, the authors discuss a secondary outcome measure. What did they conclude from this? The df for the SCL-58 test is different than for the Hamilton and Montgomery-Asberg tests. What does this tell you? How consistent are the results for the three tests? Why are the results different at all?

7. Another secondary outcome was the score on the Bech melancholia subscale. What analysis was done and what did the authors conclude? Do you agree?

8. Table 3 addresses the issue of adverse events. In the text, it is stated that 30-37% of the patients experienced some sort of adverse effect. How does this compare to the numbers in Table 3? No statistical test was done on the data in Table 3. What test would you use to determine if there was a difference in adverse effects between the two groups? What problem would arise with the usual method for most of the events? In Table 3, some patients might have experienced more than one adverse effect. What was done about this?

9. On p. 1365, it is stated that the median difference on the Hamilton score was 3.0 and the mean difference was 1.8. What, if anything, does this tell you about the distribution of scores?

10. On p. 1365, the authors state that the difference in percentage of responders was significant, but the effect size is moderate. What is a "responder?" What do the authors mean by these remarks? What is the p-value for the comparison of responders? Is this a one- or two-sided test?

Efficacy of St. John's Wort Extract WS 5570 in Major Depression: A Double-Blind, Placebo-Controlled Trial

Y. Lecrubier, M.D.

G. Clerc, M.D.

R. Didi, M.D.

M. Kieser, Ph.D.

Objective: In a double-blind, randomized, placebo-controlled trial with 375 patients the authors investigated the antidepressant efficacy and safety of 300 mg t.i.d. of hydroalcoholic *Hypericum perforatum* extract WS 5570.

Method: The study participants were male and female adult outpatients with mild to moderate major depression (single or recurrent episode, DSM-IV criteria). After a single-blind placebo run-in phase, the patients were randomly assigned, 186 to WS 5570 and 189 to placebo, after which they received double-blind treatment for 6 weeks. Follow-up visits were held after 1, 2, 4, and 6 weeks. The primary outcome measure was the change from baseline in the total score on the 17-item Hamilton Depression Rating Scale. In addition, analyses of responders (patients with at least a 50% reduction in Hamilton total score) and patients with remissions (patients with a total score of 6 or less on the Hamilton scale at treatment end)

were carried out, and subscale/subgroup analyses were conducted. The design included an adaptive interim analysis performed after random assignment of 169 patients with options for group size adjustment or early termination.

Results: Compared to placebo, WS 5570 produced a significantly greater reduction in total score on the Hamilton depression scale and significantly more patients with treatment response or remission. It was more effective in patients with higher baseline Hamilton scores and led to global reduction of depression-related core symptoms, assessed with the melancholia subscale of the Hamilton scale. The placebo and WS 5570 groups had comparable adverse events.

Conclusions: *H. perforatum* extract WS 5570 was found to be safe and more effective than placebo for the treatment of mild to moderate depression.

(Am J Psychiatry 2002; 159:1361–1366)

After decades of predominant reliance on synthetic antidepressants, the treatment of mildly and moderately severe forms of major depression with extracts from St. John's wort (*Hypericum perforatum*) is becoming increasingly popular, with sales of $86 million in the U.S. market during 2000. Today, preparations based on *H. perforatum* extract are among the most widely prescribed drugs for depression in many European countries.

The efficacy of drugs based on *H. perforatum* in alleviating mild to moderate depressive states was confirmed in comparisons with placebo and with effective synthetic standard antidepressants (e.g., imipramine, fluoxetine); reviews and meta-analyses have been conducted by Kim et al. (1), Linde et al. (2), and Volz (3). It has been claimed that *H. perforatum* is associated with fewer and less severe side effects than its active comparators (see references 1, 4–6). Despite the large body of published evidence supporting the efficacy of *H. perforatum* extract as an antidepressant, reviewers (1, 7) have identified serious design problems in existing studies and have criticized the meagerness of the database. Even in cases of a positive response with a classical scale such as the Hamilton Depression Rating Scale (8, 9), one may question whether the

observed score changes reflect a true antidepressant effect. Therefore, quantitative and qualitative data for assessing the antidepressant efficacy of *H. perforatum* extract in relation to placebo are welcome.

An aspect with potentially important clinical implications is the initial severity of the patient's depression and its relationship to treatment efficacy. Laakmann and colleagues (10) investigated mildly to moderately depressed patients with a pretreatment total score on the Hamilton depression scale (17-item version) of 17 or higher, and they suggested that antidepressant treatment with *H. perforatum* extract was more efficacious for the more severely depressed subgroup (those with an initial total score on the Hamilton scale of 22 or higher).

The aim of the present study was to compare the efficacy of *H. perforatum* extract WS 5570 to that of placebo in a large group of patients suffering from a mild to moderate major depressive episode according to DSM-IV. In addition, particular attention was paid to the efficacy observed for the core symptoms of depression, as measured by the Bech melancholia scale (11), and to the relationship between the initial severity of depression and response to treatment.

Method

This was a 6-week double-blind, placebo-controlled, randomized phase III trial comparing the efficacy of WS 5570, 300 mg t.i.d., and placebo. The investigation was conducted by the Hypericum Study Group between July 1997 and June 2000 in 26 clinical centers in France. The European Union's Good Clinical Practice guidelines, the Declaration of Helsinki, and national regulatory and legal requirements (French Code of Public Health), including approval of the trial protocol by an independent ethics committee, were observed. After complete description of the study, written informed consent was obtained from all subjects.

Centers and Subjects

The study participants were recruited from the pool of patients who sought treatment for depression in any of the clinical centers and who met the trial's entry criteria. Most centers were outpatient departments associated with psychiatric inpatient departments, while other centers were situated in private practices. Patient inclusion and all evaluations of ratings were conducted by psychiatrists. All investigators participated in specific training to identify and include patients with the appropriate diagnosis. They were trained to use a structured interview, the Mini-International Neuropsychiatric Interview (12, 13). Standardization of ratings was ensured by the rating of videotaped patient interviews by all investigators and subsequent discussion of appropriate ratings. One videotaped patient was at the upper range of severity (mean total score on Hamilton depression scale=26.3, SD=2.0); a second patient was at the lower end of the range (mean score= 20.8, SD=1.5). None of the selected investigators showed a rating deviating by more than two standard deviations from the mean rating.

A patient was eligible for the study if he or she 1) was an outpatient aged 18 to 65 years at the time of screening, 2) provided written informed consent, 3) had a current major depressive episode of at least 2 weeks' duration that met the criteria of DSM-IV code 296.21, 296.22, 296.31, or 296.32 (mild or moderate depression, single or recurrent episode), and 4) had a total score on the Hamilton depression scale between 18 and 25 and a score on item 1 ("depressed mood") of 2 or higher at screening and baseline. The reasons for exclusion were depression of any type other than those specified, any serious psychiatric disease other than depression, serious suicidal risk (score of 3 or higher on item 3 of the Hamilton depression scale), or response to placebo during the run-in phase; response was defined as a 25% or greater reduction of the Hamilton depression scale total score.

Investigational Treatments

WS 5570 is a hydroalcoholic extract from Herba hyperici (drug-to-extract ratio, 4–7:1) with standardized contents of 3%–6% hyperforin and 0.12%–0.28% hypericin according to high-performance liquid chromatography. The drug was presented in film-coated tablets, each of which contained 300 mg of the extract.

The tablets containing placebo were indistinguishable from those containing WS 5570 in all aspects of their outward appearance.

Procedure

After giving written informed consent, the patients underwent a screening examination to determine their eligibility for the trial and entered a single-blind placebo run-in period of 3 days for patients who did not need a wash-out and at least 7 days when medications had to be withdrawn before randomized treatment. During a baseline examination (day 0) with reassessment of the entry criteria, eligible patients were randomly assigned at a ratio of 1:1 to treatment with 300 mg t.i.d. of H. perforatum extract WS 5570 or placebo administered over 6 weeks (14). Efficacy and safety were evaluated after 7, 14, 28, and 42 days of randomized treatment.

Measures of Efficacy and Safety

The primary outcome measure for treatment efficacy was the change in the total score on the Hamilton depression scale (17-item version) in the intention-to-treat data set between baseline (day 0) and subsequent visits during randomized treatment. The reliability and validity of the Hamilton depression scale for this population have been previously demonstrated (15). Secondary measures of efficacy were the total score on the Montgomery-Åsberg Depression Rating Scale (16), the score on the 58-item version of the Symptom Check List (SCL-58) (17), and the Clinical Global Impression (18).

In addition to the total score on the Hamilton depression scale, the melancholia subscore was analyzed. The melancholia subscale was derived from work by Bech and colleagues (11), who applied three formal psychometric criteria (calibration, ascending monotonicity, dispersion) to each item of the Hamilton depression scale. The additive combination of the six items that fulfilled all three criteria was suggested by the authors as a "valid subscale" of the Hamilton depression scale that primarily includes the items that measure the core symptoms of depression. The secondary analysis of the Hamilton scale also included an assessment of responder rates, which were determined as the percentage of patients in each treatment group whose total score on the Hamilton depression scale at the end of treatment was at least 50% lower than at baseline.

Safety measures comprised physical examinations and laboratory tests before and after double-blind treatment (glucose, sodium, potassium, aspartate transaminase, alanine transaminase, γ-glutamyltransferase, serum creatinine, thyrotropin, hemoglobin, hematocrit, RBC count, WBC count and differential, platelet count). Vital signs were tested at each of the visits. In addition, the patients were thoroughly questioned for adverse events in a general inquiry during all follow-up visits.

Statistical Analysis

Previous investigations of mild to moderate depression have shown that the extent of the response to placebo treatment is hardly predictable in advance and tends to be very variable; for instance, in the 13 placebo-controlled studies reviewed by Linde et al. (2), the rates of response with placebo treatment ranged from 0% to 54%. In addition, the studies also showed striking differences regarding the magnitude of the within-group variability of the response to treatment. To avoid the ethically questionable exposure of an unnecessarily large number of depressed patients to placebo and to minimize the risk of insufficient statistical power, we planned and conducted our study with an adaptive interim analysis (19). Depending on the results of the interim analysis, the study would stop with early rejection or acceptance of the null hypotheses to be tested or it would continue with a second stage in which the number of subjects would be determined by using the results of the first study part. The overall significance level for confirmatory analysis was $\alpha=0.025$, one-sided, according to guideline E9 of the International Conference on Harmonization (20) (this corresponds to a two-sided level of $\alpha=0.05$). The boundaries for early rejection or acceptance of a null hypothesis in the interim analysis were $\alpha_1=0.0153$ or $\alpha_0=0.20$ (both one-sided). If the one-sided p value of the first stage, p_1, lay between these boundaries ($\alpha_1<p_1<\alpha_0$), the study would continue with a second part. The null hypothesis could then be rejected if $p_1 \cdot p_2 \leq 0.0038$, where p_2 is the one-sided p value determined from the second stage of the study. These boundaries were computed from the methodology given by Bauer and Koehne (19). In the confirmatory analysis, comparisons of treatment groups were performed for the differences in Hamilton depression scale total score between baseline and day 42, day 28, day 14, and day 7. A null hypothesis was tested only if the results of the preceding tests were significant so that the experi-

TABLE 1. Baseline Characteristics of Patients With Major Depression Who Were Treated With *Hypericum Perforatum* Extract WS 5570 or Placebo

Characteristic	Patients Included in the Intention-to-Treat Analysis					
	WS 5570 (N=186)			Placebo (N=189)		
	N	%		N	%	
Sex						
Male	44	23.7		44	23.3	
Female	142	76.3		145	76.7	
Severity of illness according to Clinical Global Impression						
Mildly ill	25	13.4		21	11.1	
Moderately ill	97	52.2		111	58.7	
Severely ill	64	34.4		57	30.2	
	Mean	SD	Range	Mean	SD	Range
Age (years)	40.2	11.7	18.9–65.5	41.2	11.4	18.6–66.0
Hamilton Depression Rating Scale (17-item) total score	21.9	1.7	18–27	21.9	1.7	18–25
Montgomery-Åsberg Depression Rating Scale total score	24.2	3.9	14–34	24.5	4.2	8–34
SCL-58 total score	64.3	25.9	9–160	68.4	29.3	18–186

ment-wise type I error rate would be controlled without an adjustment of the significance level (21).

The hypotheses were evaluated by using two-sample t tests. All other analyses were purely descriptive without adjustment for multiplicity. For the assessment of the course of change in Hamilton depression score during treatment, a repeated measures analysis of variance (ANOVA) was applied. To investigate the relationship between treatment efficacy and the severity of depression before the start of treatment, an explorative subgroup analysis was conducted for patients with an initial Hamilton depression scale total score of less than 22 and for those with a score of 22 or higher.

The primary analysis of efficacy was based on the intention-to-treat principle and included all randomly assigned patients. An additional per-protocol analysis of the patients without major protocol deviations was conducted for the primary outcome measure. For all efficacy analyses, the last observation was carried forward for patients who terminated the trial prematurely. Safety analyses were based on all patients who took at least one dose of the study medication after random assignment.

Results

Patient Accountability

In the first part of the trial, 169 patients were randomly assigned to double-blind treatment (WS 5570: N=84; placebo: N=85), and 206 patients were randomly assigned in the second part after the interim analysis (WS 5570: N=102; placebo: N=104). Therefore, totals of 186 and 189 patients were randomly assigned to treatment with WS 5570 and placebo, respectively, and were included in the intention-to-treat analysis. After random assignment, 18 patients in the WS 5570 group (9.7%) and 25 in the placebo group (13.2%) terminated treatment prematurely. The primary reasons for early withdrawal were lack of efficacy (WS 5570: N=10, 5.4%; placebo: N=14, 7.4%), revocation of informed consent (WS 5570: N=4, 2.2%; placebo: N=7, 3.7%), and adverse events (WS 5570: N=2, 1.1%; placebo: N=2, 1.1%). The per-protocol analysis of all patients without major protocol violations included 164 patients in the WS 5570 group and 157 patients in the placebo group. The decisions with respect to the relevance of the protocol deviations were made before the code was broken.

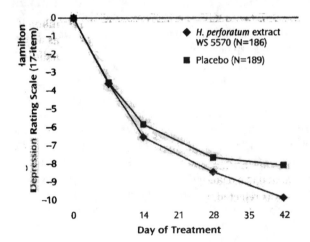

FIGURE 1. Change in Score on Hamilton Depression Rating Scale in Patients With Major Depression During 6 Weeks of Treatment With *Hypericum Perforatum* Extract WS 5570 or Placebo

Baseline Characteristics

Basic demographic characteristics and ratings of pretreatment severity of illness for the patients included in the intention-to-treat analysis are presented in Table 1. The data show that the two treatment groups were essentially comparable at baseline. In particular, there were no relevant differences regarding the patients' severity of depression according to the total scores on the Hamilton depression scale and the Montgomery-Åsberg Depression Rating Scale.

Efficacy

Between the start of randomized treatment and day 42, both groups' average total scores on the Hamilton depression scale decreased monotonically (Figure 1). Starting from baseline mean values of 21.9 points (SD=1.7) in both treatment groups, the Hamilton depression total score decreased during the treatment phase by a mean of 9.9 points (SD=6.8) in the WS 5570 group and by 8.1 points

TABLE 2. Relation of Baseline Score on Hamilton Depression Rating Scale to Score Change and to Responder Rate for Patients With Major Depression Who Were Treated for 6 Weeks With *Hypericum Perforatum* Extract WS 5570 or Placebo

Baseline Total Score on Hamilton Scale (17-item) and Outcome Variable	Patients in Intention-to-Treat Analysis							
	WS 5570 (N=186)				Placebo (N=189)			
	N	%	Mean	SD	N	%	Mean	SD
Baseline score <22	76	—			72	—		
Responders	38	50.0			30	41.7		
Total score change			−9.0	6.3			−7.4	6.0
Baseline score ≥22	110	—			117	—		
Responders	60	54.5			50	42.7		
Total score change[a]			−10.5	7.0			−8.5	7.7
Total data set	—				—			
Responders	98	52.7			80	42.3		
Total score change[b]			−9.9	6.8			−8.1	7.1

[a] Significantly greater reduction in the WS 5570 group (t=2.08, df=225, p=0.04, two-sided t test).
[b] Significantly greater reduction in the WS 5570 group (t=2.50, df=373, p=0.03, two-sided t test).

(SD=7.1) in the placebo group (pooled data from both study stages; last observation carried forward).

In the confirmatory hypothesis testing for the primary outcome measure for the first study stage, i.e., the interim analysis for the intention-to-treat data set, the null hypothesis relating to the difference between treatment groups in the decrease in total score on the Hamilton depression scale between baseline and day 42 was associated with a one-sided p value of $p_1=0.037$ (t=1.80, df=167). Since this p value lies between the boundaries for early rejection and acceptance, the trial was continued with a second stage. The required number of subjects was reestimated on the basis of the results of the interim analysis. The group in the second stage showed a one-sided p value of $p_2=0.038$ (t=1.78, df=204). Therefore, the product of p values for the final combination test fell below the critical limit (0.037·0.038=0.0014<0.0038), and so the null hypothesis was rejected, and the superiority of extract WS 5570 over placebo was demonstrated for a treatment duration of 6 weeks. For the pooled data from both study stages the one-sided p value for the change in Hamilton depression score between baseline and day 42 was p=0.02 (t=2.50, df=373). For the comparisons of the two treatment groups in terms of change from baseline in Hamilton scale total score at days 28, 14, and 7, the t test results were nonsignificant. A repeated measures ANOVA with independent variables of treatment and time and an interaction term was used to compare the postbaseline Hamilton depression scores of the two treatment groups and demonstrated a significant time-by-treatment interaction (F=3.41, df=4, 1492; Greenhouse-Geisser epsilon=0.58, two-sided Greenhouse-Geisser-corrected p=0.03).

These results were confirmed in the per-protocol analysis, in which both treatment groups had the same mean decreases in Hamilton depression scale total score between baseline and day 42 as in the intention-to-treat analysis (t=2.31, df=319, p=0.02, two-sided t test).

In the intention-to-treat study group, the percentage of responders (those with at least 50% decreases in Hamilton score between baseline and treatment end) was significantly higher for WS 5570 (52.7%, 98 of 186) than for placebo (42.3%, 80 of 189) (χ^2=4.04, df=1, p<0.05, two-sided). Furthermore, the percentage of patients with remission (score of 6 or less on Hamilton scale at treatment end) was significantly higher for the active treatment group (24.7%, 46 of 186) than for placebo (15.9%, 30 of 189) (χ^2=4.55, df= 1, p=0.03, two-sided).

A secondary outcome measure was the change in total score on the Montgomery-Åsberg Depression Rating Scale between baseline and treatment end. The mean decrease was 11.7 points (SD=9.0) for the WS 5570 group and 9.9 points (SD=9.2) for the placebo group (intention-to-treat analysis: t=1.90, df=373, p=0.06, two-sided t test). The depression subscore of the SCL-58 (11 items) showed a mean reduction of 7.9 points (SD=8.7) for the WS 5570 group and 6.5 points (SD=8.4) for the placebo group (intention-to-treat analysis: t=1.57, df=366, p=0.12, two-sided t test).

Table 2 indicates the relationship between the initial severity of depression and the magnitude of the treatment effect. Among the patients receiving WS 5570, the difference in the decrease in Hamilton depression scale total score between baseline and the final visit was larger in the subgroup of patients with initial scores equal to or above the median value of 22 points. Their decrease was significantly greater than the decrease for the patients receiving placebo (t=2.08, df=225, p=0.04, two-sided t test), but the decrease for the patients with initial Hamilton depression scores between 18 and 21 did not differ significantly from that for the placebo group (t=1.50, df=146, p=0.14, two-sided t test).

The score on the Bech melancholia subscale decreased by a mean of 5.5 points (SD=4.2) in the WS 5570 group and by 4.4 points (SD=4.1) in the placebo group (intention-to-treat analysis: t=2.60, df=373, p=0.001, two-sided t test).

Safety and Tolerability

During double-blind treatment with WS 5570, 30.6% of the patients (57 of 186) experienced adverse events, compared to 37.0% (70 of 189) in the placebo group. The type, incidence (Table 3), and severity of adverse events did not indicate any treatment-emergent risks associated with WS 5570. In each study group, two patients were withdrawn prematurely because of adverse events. In the WS 5570 group, both withdrawals were necessitated by hospitalization of the patient because of symptom aggravation.

Both study drugs did not relevantly or systematically modify the biological measures assessed—neither with regard to a general trend nor on an individual patient basis.

Discussion

This study demonstrates the antidepressant efficacy of 300 t.i.d. of *H. perforatum* extract WS 5570, as compared to placebo, for mildly to moderately depressed patients after a treatment duration of 6 weeks.

The median difference in change in the Hamilton depression scale score between WS 5570 and placebo was 3.0 points, and the mean difference was 1.8 points. The percentage of responders in the group receiving *H. perforatum* was 52.7%, whereas the percentage in the placebo group was 42.3%; the difference of 10.4% was significant. This effect size is moderate, but the size of the effect observed in placebo-controlled phase III trials hardly reflects the real therapeutic potential. As observed by Montgomery (22), during recent years an increasing number of clinical trials testing the efficacy of new potential antidepressants have failed to demonstrate a difference from placebo even for established reference drugs (e.g., imipramine). The rise in placebo response rates has not been paralleled by a rise in drug response. Similar tendencies have been observed in trials for other psychiatric indications (e.g., obsessive-compulsive disorder, social phobia, panic disorder).

Placebo-controlled phase III trials control for different sources of bias, making the therapeutic management artificial, e.g., no associated treatment in case of anxiety or insomnia, no dose adaptation depending on tolerance and efficacy. These methodological constraints are justified but likely to decrease the effect size of the active drug. On the other hand, the effect of the "therapeutic alliance" with the physician, including confidence, empathy, and helpful attitude, is increased by the awareness of the possible random assignment to a new innovative drug and the existence of weekly visits, thus increasing the placebo effect (23, 24). One of the strategies proposed to increase the difference between active drug and placebo was to recruit rather severely depressed patients (e.g., having a Hamilton depression scale score above 25), although it is not clear whether such a threshold effect is due to a limited placebo response in severely ill patients or to a mechanical effect yielding a larger difference between the evaluations before and after treatment. Therefore, the superiority observed in this study of mildly and moderately depressed patients was probably more difficult to evidence than it would have been with a group including more severely ill patients. In a recent well-designed and carefully conducted study (25), *H. perforatum* was compared to placebo and to a reference drug, and a significant effect was not demonstrated

TABLE 3. Number of Patients With Major Depression Who Experienced Adverse Events During 6 Weeks of Treatment With *Hypericum Perforatum* Extract WS 5570 or Placebo[a]

| | Patients in Intention-to-Treat Analysis | | | |
| | WS 5570 (N=186) | | Placebo (N=189) | |
Adverse Event	N	%	N	%
Nausea	9	4.8	6	3.2
Headache	3	1.6	7	3.7
Dizziness	4	2.2	4	2.1
Abdominal pain	2	1.1	4	2.1
Insomnia	3	1.6	2	1.1

[a] Adverse events for which a causal relationship with the study drug could not be excluded and that were observed in at least three patients in one treatment group.

even for the reference compound. After reviewing the methods of the two studies, we consider only two major differences potentially relevant: our study had a small number of patients at each center, but all patients recruited were spontaneously seeking care. The difference in care-seeking behavior, reflecting a subjective perception of severity by the patient, is likely to explain the better detection of an advantage of active drug over placebo. A low dropout rate, possibly due to a low rate of adverse events during the first weeks of treatment, may have contributed to this positive result. The existence of a run-in phase was obviously of limited value, as already claimed by others (22, 26, 27). However, as suggested early on by Paykel et al. (28), we observed that severity did have an impact on outcome. The difference between treatment groups in the change from baseline in Hamilton depression score was significant for the subgroup of patients with baseline scores of 22 or higher but not for those with baseline scores less than 22, confirming the results of Laakmann et al. (10). These authors found in the *H. perforatum* group (extract WS 5572) responder rates of 50% for those with baseline Hamilton depression scores of 22 or higher and 49% for all patients, compared to 25% and 33%, respectively, in the placebo group.

The significant effect on the Bech melancholia subscale score and a larger effect in more severely ill patients suggest a true antidepressant effect of *H. perforatum*. The results observed with the Montgomery-Åsberg Depression Rating Scale were consistent with those of the Hamilton depression scale and shortly missed significance (p=0.06).

The duration of the study was 6 weeks because this duration was recommended for placebo-controlled phase III studies of acutely depressed patients. Greater improvement in patients receiving active drug than in those receiving placebo is commonly observed thereafter. Indeed, the improvement shown by the *H. perforatum* group in Figure 1 has not yet reached a plateau.

In the present trial, the two withdrawals related to adverse events in the *H. perforatum* group were attributable to worsening depressive symptoms due to a lack of efficacy in these particular patients, rather than to problems regarding tolerability. The herbal extract was well tolerated by all 186 patients who received it, and tolerability-related withdrawal from treatment was not indicated in a single case. Our data therefore contribute to the overall favorable safety assessment of preparations from *H. perforatum* extract.

In conclusion, this study demonstrates the existence of an antidepressant effect of *H. perforatum* in mildly and moderately depressed patients.

Received July 26, 2001; revision received April 4, 2002; accepted March 27, 2002. From the Unité Institut National de la Santé et de la Recherche Médicale 302, Hôpital Pitié Salpêtrière; the Centre Hospitalier Spécialisé en Psychiatrie de Pontorson, Pontorson, France; the Centre Hospitalier Spécialisé en Psychiatrie de La Chartreuse, Dijon, France; and the Biometrical Department, Dr. Willmar Schwabe Pharmaceuticals, Karlsruhe, Germany. Address reprint requests to Dr. Lecrubier, Unité INSERM 302, Hôpital Pitié Salpêtrière, Pavillon Clérambault, 75013 Paris, France; lecru@ext.jussieu.fr (e-mail).

Sponsored by Dr. Willmar Schwabe Pharmaceuticals, Karlsruhe, Germany.

Hypericum Study Group: D. Arnoux, G. Aspe, J.J. Ausseill, C. Bagot, A. Benoit, P. Bern, D. Bonnaffoux, B. Bonnet Guerin, J. Charbaut, J.Y. Charlot, C. Claden, G. Clerc, H. de Verbigier, R. Didi, M. Faure, F. Gheysen, S. Guibert, P. Khalifa, P. Le Goubey, D. Liegaut, Y. Mouhot, A. Pargade Moradell, L. Rochard, G. Saint Mard, H. Sauret, and J.R. Zekri.

References

1. Kim HL, Streltzer J, Goebert D: St John's wort for depression—a meta-analysis of well-defined clinical trials. J Nerv Ment Dis 1999; 187:532–539
2. Linde K, Ramirez G, Mulrow CD, Pauls A, Weidenhammer W, Melchart D: St John's wort for depression—an overview and meta-analysis of randomized clinical trials. Br Med J 1996; 313: 253–258
3. Volz HP: Controlled clinical trials of Hypericum extracts in depressed patients—an overview. Pharmacopsychiatry 1997; 30(suppl 2):72–76
4. Ernst E: St John's wort, an anti-depressant? a systematic, criteria-based review. Phytomedicine 1995; 2:67–71
5. De Smet PAGM, Nolen WA: St John's wort as an antidepressant. Br Med J 1996; 313:241–242
6. Ernst E, Rand JI, Barnes J, Stevinson C: Adverse effects profile of the herbal antidepressant St John's wort (Hypericum perforatum L). Eur J Clin Pharmacol 1998; 54:589–594
7. Deltito J, Beyer D: The scientific, quasi-scientific and popular literature on the use of St John's wort in the treatment of depression. J Affect Disord 1998; 51:345–351
8. Hamilton M: A rating scale for depression. J Neurol Neurosurg Psychiatry 1960; 23:56–62
9. Hamilton M: Development of a rating scale for primary depressive illness. Br J Soc Clin Psychol 1967; 6:278–296
10. Laakmann G, Dienel A, Kieser M: Clinical significance of hyperforin for the efficacy of Hypericum extracts on depressive disorders of different severities. Phytomedicine 1998; 5:435–442
11. Bech P, Gram LF, Dein E, Jacobsen O, Vitger J, Bowling TG: Quantitative rating of depressive states. Acta Psychiatr Scand 1975; 51:161–170
12. Lecrubier Y, Sheehan DV, Weiller E, Amorim P, Bonara I, Sheehan KH, Janavs J, Dunbar GC: The Mini-International Neuropsychiatric Interview (MINI), a short diagnostic structured interview: reliability and validity according to the CIDI. Eur Psychiatry 1997; 12:224–231
13. Sheehan DV, Lecrubier Y, Sheehan KH, Amorim P, Janavs J, Weiller E, Hergueta T, Baker R, Dunbar GC: The Mini-International Neuropsychiatric Interview (MINI): the development and validation of a structured diagnostic psychiatric interview for DSM-IV and ICD-10. J Clin Psychiatry 1998; 59(suppl 20):22–33
14. Committee for Proprietary Medicinal Products: Note for Guidance on Clinical Investigation of Medicinal Products in the Treatment of Depression (draft). London, EMEA: The European Agency for the Evaluation of Medicinal Products, 2001
15. Hedlund JL, Vieweg BW: The Hamilton Rating Scale for Depression: a comprehensive review. J Operational Psychiatry 1979; 10:149–165
16. Montgomery SA, Åsberg M: A new depression scale designed to be sensitive to change. Br J Psychiatry 1979; 134:382–389
17. Guelfi JD, Barthelet G, Lancrenon S, Fermanian J: Structure factorielle de la HSCL: sur un échantillon de patients anxio-dépressifs français. Ann Med Psychol 1984; 142:889–896
18. Guy W (ed): ECDEU Assessment Manual for Psychopharmacology: Publication ADM 76-338. Washington, DC, US Department of Health, Education, and Welfare, 1976, pp 218–222
19. Bauer P, Koehne K: Evaluation of experiments with adaptive interim analyses. Biometrics 1994; 50:1029–1041
20. International Conference on Harmonization: Note for Guidance on Statistical Principles for Clinical Trials: ICH Topic E9. London, EMEA: The European Agency for the Evaluation of Medicinal Products, 1998
21. Kieser M, Bauer P, Lehmacher W: Inference on multiple endpoints in clinical trials with adaptive interim analyses. Biometrical J 1999; 41:261–277
22. Montgomery SA: The failure of placebo-controlled studies. Eur Neuropsychopharmacol 1999; 9:271–276
23. Montgomery SA: Alternatives to placebo-controlled trials in psychiatry. Eur Neuropsychopharmacol 1999; 9:265–269
24. Schweizer E, Rickels K: Placebo response in generalized anxiety: its effect on the outcome of clinical trials. J Clin Psychiatry 1997; 58(suppl 11):30–38
25. Davidson J: The NIH Hypericum study in depression, in Proceedings of the 40th Annual Congress of the American College of Neuropsychopharmacology. Nashville, Tenn, ACNP, 2001, p 23
26. Greenberg RP, Fisher S, Riter JA: Placebo washout is not a meaningful part of antidepressant drug trials. Percept Mot Skills 1995; 81:688–690
27. Bialik RJ, Ravindran AV, Bakish D, Lapierre YD: A comparison of placebo responders and nonresponders in subgroups of depressive disorder. J Psychiatry Neurosci 1995; 20:265–270
28. Paykel ES, Hollyman JA, Freeling P, Sedgewick P: Predictors of therapeutic benefit from amitriptyline in mild depression: a general practice placebo-controlled trial. J Affect Disord 1988; 14:83–95

"Biology of a deep benthopelagic fish, roudi excolar, *Promethichthys promestheus* (Gempylidae), off the Canary Islands," Jose M. Lorenzo, Jose G. Pajeulo, *Fishery Bulletin*, 97(1) 1999, pp. 92-99.

Background
This article discusses characteristics of a particular species of fish caught off the Canary Islands.

Questions
1. What variables were measured in the study? Which are categorical and which are quantitative?
2. Figure 2 is the "length-frequency distribution." What is the name for this type of display? What does this tell you about the length of the fish that were caught? Does the length distribution appear to be normal? If not, is the difference from normal important?
3. Tables 1, 2, and 3 give the sex distribution vs. different variables. The authors use the χ^2 statistic. If you have not covered this statistic, how could you compare the sex distribution? If you square your statistic, you should get the χ^2 that is given. State H0 and Ha. Why is the statistic so much larger for Total than for any single length category? Does the statistical test tell you something that was not obvious from the data?
4. How would you test if the sex ratio was the same for all length categories? (This is not what the authors did.)
5. Table 2 gives the sex ratios for each quarter of the year. What did the researchers find regarding the relationship between sex ratios and time (quarter)? Suppose we combine all the data for quarters 2 and 3 and do the analysis. Repeat, combining quarters 1 and 4. Does this change the conclusion? If so, why did this happen? Is it permissible to combine the quarters the way we did? That is, is it permissible to combine quarters that appear to be similar and then performing the statistical test? (The answer to this question is quite complex. Basically, it depends on what you are willing to assume.)
6. Table 3 gives the sex ratios by depth. What did the authors conclude about this relationship? Calculate the overall χ^2 statistic. What do you conclude from this statistic? How does this compare to what the authors concluded? There is a technique called "pooling" that allows you to start with a large table (seven rows in this case) and collapse it into a smaller table that is easier to interpret. We can combine rows (in this case) and re-calculate the statistic. The only adjustment required is that the df remains at its original value. Try pooling rows for depths 201-500 and pooling rows for depths 601-900. Leave 501-600 alone because it doesn't seem to match either the shallower depths or the deeper depths. What would you conclude from this statistic?
7. What type of graph is Figure 3? What do you think the vertical lines on the graph mean? What do the authors conclude from this graph? Do you agree? What statistical evidence did the authors cite (if any)?
8. Table 4 gives a and b for the length-weight relationship. What are these in terms of your statistics course? Which variable is X and which is Y? Does it matter? Of a and b, which is the slope and which is the intercept? (Not everyone uses the same

notation.) The authors give a value for "t test" in Table 4. Can you confirm this value? (I cannot.) What conclusion would you draw from the given information?

9. The authors then go on to compare the two slopes to see if the relationship for males is the same as for females. This could be done by taking the difference in the two slopes. How would you calculate the SE for this difference from the values in Table 4. What would df be for the difference? Do you need an exact value for df? Explain.

10. Consider Figure 5. The vertical axis is labeled "Frequency." Is this a histogram? If so, why is it displayed as lines instead of the traditional bars? What do the authors conclude from this figure? Do you agree? What statistical evidence do they cite? What statistical evidence would you have given?

11. Table 5 is titled "Age-length key." What do these numbers represent? That is, the row labeled 36 and the column labeled III contain a "2." What does this mean? The bottom of the table gives some statistics for each column. Can you confirm these values? (We presume that the second row, labeled "x" is actually the average.) Describe the length distribution within each age group. As fish get older, they get longer. Does the variation in length seem to change with age? Is this reasonable? How would you do a statistical test to see if length has a normal distribution for a given age group? How special problems are caused by the fact that the shortest length category has only a few fish in each age category?

Biology of a deep benthopelagic fish, roudi escolar *Promethichthys prometheus* (Gempylidae), off the Canary Islands

José M. Lorenzo
José G. Pajuelo
Departamento de Biología (Universidad de Las Palmas de Gran Canaria)
Edificio de Ciencias Básicas, Campus Universitario de Tafira
35017 Las Palmas, Spain
E-mail address (for J. M. Lorenzo): josemaria.lorenzo@biologia.ulpgc.es

Abstract—The roudi escolar *Promethichthys prometheus* is common in deep hook-and-line and longline catches of a small-scale fishery along the slope off the Canary Islands. Population structure, reproduction, growth, and mortality of the species were studied from sampling undertaken from August 1992 to July 1995. Range of length of fish in the catches was between 36 and 80 cm TL, with a main distribution between 56 and 66 cm. The overall ratio of males to females was 1:1.74. Females predominated in all sizes. The sex ratio varied throughout the period of study; the lowest discrepancy between males and females, however, was during the reproductive period. A vertical space partitioning among sexes was observed, with males predominating from 600 to 800 m depth, females from 300 to 500 m. The reproductive period of the species was from April to September, with a peak in spawning in June–July. The size at first maturity was 47.41 cm. The parameters of the length-weight relationship for all fish were a=0.004521 and b=2.98932. Age readings of otoliths indicated that the exploited population consisted of nine age groups (III–XI years). The von Bertalanffy growth parameters for all individuals were L_∞=93.61 cm, k=0.18/years, and t_0 =−1.54 years. The rates of mortality for all fish were Z=0.49/years, M=0.35/years, and F=0.14/years. The length at first capture for the whole population was 51.57 cm.

Manuscript accepted 23 April 1998.
Fish. Bull 97:92–99 (1999).

The family Gempylidae consists of 16 genera and 23 species. Only seven species are found off the Canary Islands, one of which is the roudi escolar *Promethichthys prometheus* (Cuvier, 1832), the only species recognized to date in the genus *Promethichthys* (Nakamura and Parin, 1993).

The roudi escolar is a benthopelagic marine fish that has a worldwide distribution in tropical and warm temperate waters. This species generally inhabits waters between 100 and 800 m in depth over seamounts and continental and insular slopes. It migrates upward at night, probably forming schools (Nakamura, 1981; Parin, 1986; Nakamura and Parin, 1993).

Published information on *P. prometheus* is very scarce. The majority of studies describe its morphological characteristics, geographical and depth distribution, and ecology (Nakamura, 1981; Parin, 1986; Nishikawa, 1987; Nakamura and Parin, 1993). Only Lorenzo and Pajuelo (1995) have studied some biological aspects of the species. These authors carried out a preliminary study on the sex ratio, reproduction, and age and growth of roudi escolar off the Canary Islands (central-east Atlantic) on the basis of a small number of specimens during one life cycle. This paper is an extension of their work, analyzing, in addition to all those aspects, population structure and mortality.

The roudi escolar is common in the catches of the deep hook-and-line and longline small-scale fishery over the slope off the Canary Islands. In this area, this species is captured year round without significant seasonal differences in landings.

Materials and methods

Between August 1992 and July 1995, the TL (cm) of 1879 specimens of roudi escolar was measured monthly from commercial catches of the small-scale fleet. Fish were caught with baited hook-and-lines and longlines at depths of 285–870 m around the islands of the Canary archipelago (Fig. 1).

A subsample was taken by a random stratified method from each sample for biological examination. In total, 776 individuals were analyzed. For each fish, the TW (0.1 g) and the weight of the gonads (0.01 g) were measured, and sex and stage of maturation were ascertained macroscopically. The latter was classified as follows: I = immature; II = resting; III = ripe; IV = ripe and running; V = spent. Sagittal otoliths of the fish were extracted, cleaned, and stored dry. The length-frequency distribution of individuals in catches was calculated. Data were pooled for 1992–95.

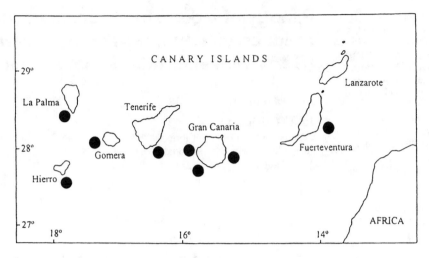

Figure 1
Location of sampling areas (●) in the Canary Islands.

The sex ratios (males:females) for the whole sample, for different size classes, for quarters of a year, and for depth strata were calculated. The reproductive season was determined on the basis of monthly variations of the gonadosomatic index (GSI) according to Anderson and Gutreuter (1983). The length at sexual maturity (length at which fifty percent of the specimens became mature) was estimated by means of a logistic function that was fitted to the proportion of the mature individuals (stages III, IV, and V) by using a nonlinear regression (Saila et al., 1988).

The ratio of total length to total weight was calculated over the whole period for males and females separately, as well as for the population as a whole, by applying a linear regression (Ricker, 1973). Age was determined by interpreting growth rings on the otoliths; whole otoliths were placed in a watch glass with a blackened bottom and containing glycerin and examined under a compound microscope with reflected light. Counts for each specimen were performed at least twice and only coincident readings were accepted. An index of average percent error (APE) developed by Beamish and Fournier (1981) was used to compare the precision of age determinations. Ageing was validated indirectly by examination of monthly changes in appearance of the margins of the otoliths (Morales-Nin, 1987). The date 1 July was considered as birthdate to assign the individual ages to age groups. The von Bertalanffy growth curve was fitted to data of the resulting age-length key by means of the Marquardt's algorithm for nonlinear least squares parameter estimation (Saila et al., 1988).

Length-frequency data were converted to age frequencies by using the estimated von Bertalanffy growth parameters (Pauly, 1983, 1984). The rate of total mortality (Z) was calculated from the length converted catch curve by using ELEFAN program (Gayanilo et al., 1988). The rate of natural mortality (M) was determined from the equation of Pauly (1980). Following estimation of Z and M, the rate of fishing mortality (F) was calculated by substraction. The length at first capture was estimated from the selection ogive generated from the length converted catch curve (Pauly, 1984).

Results

The size-frequency distribution showed a length range of 36 to 80 cm TL in the catches, with a main distribution between 56 and 66 cm (Fig. 2).

Of the 776 fish examined, 282 (36.3%) were male, 491 (63.3%) female. The sex of the remaining 3 (0.4%) individuals could not be identified macroscopically because they were immature and had very thin, translucent gonads. The overall ratio of males to females was 1:1.74 and χ^2 analysis revealed this to be significantly different from a 1:1 ratio (Table 1). Females predominated in all size intervals. Sex ratios for males and females grouped into 5-cm length classes had significant departures from the 1:1 ratio for all size intervals (Table 1). The ratio of males to females varied throughout the period of study, but there were no significant differences from the 1:1 ratio during the spring and summer months (Table 2). There was a relationship between the sex of roudi escolar and depth; males predominated at 600 to 800 m depths, females at 300 to 500 m (Table 3).

The GSI showed higher values for females than for males (Fig. 3). The same temporal variation pat-

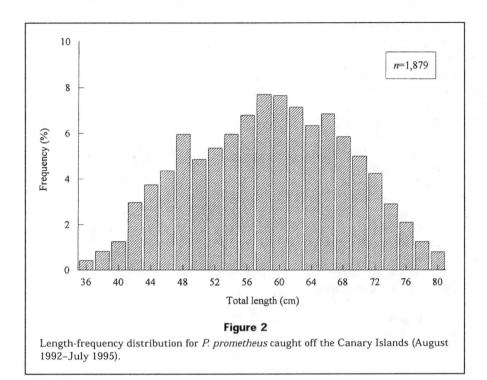

Figure 2

Length-frequency distribution for *P. prometheus* caught off the Canary Islands (August 1992–July 1995).

Table 1

Number of males and females of *P. prometheus* caught off the Canary Islands by 5-cm size class and sex ratio tested by chi-square analysis. $*=\chi^2>\chi^2_{t\,1,0.05}=3.84$.

Length (cm)	Males	Females	Sex ratio	χ^2
35.1–40.0	5	14	1:2.80	4.26*
40.1–45.0	21	40	1:1.90	5.91*
45.1–50.0	29	52	1:1.79	6.53*
50.1–55.0	45	71	1:1.57	5.82*
55.1–60.0	60	94	1.1.56	7.50*
60.1–65.0	48	84	1:1.75	9.80*
65.1–70.0	36	62	1:1.72	6.89*
70.1–75.0	26	48	1:1.76	6.54*
75.1–80.0	12	26	1:2.16	5.15*
Total	282	491	1:1.74	56.50*

Table 2

Number of males and females of *P. prometheus* caught off the Canary Islands by quarter and sex ratio tested by chi-square analysis. $*=\chi^2>\chi^2_{t\,1,0.05}=3.84$.

Quarter	Males	Females	Sex ratio	χ^2
3/92	16	23	1:1.44	1.25
4/92	13	35	1:2.69	10.08*
1/93	13	37	1:2.85	11.52*
2/93	24	37	1:1.54	2.77
3/93	24	34	1:1.42	1.72
4/93	14	38	1:2.71	11.07*
1/94	29	51	1:1.75	6.05*
2/94	36	49	1:1.36	1.98
3/94	39	50	1:1.28	1.35
4/94	24	44	1:1.83	5.88*
1/95	31	59	1:1.90	8.71*
2/95	19	34	1:1.79	4.24*

tern was recorded for both sexes. Highest values occurred between April and September, peaking during June–July. From October to March the values were low.

No significant difference in length at sexual maturity was found between males and females (*t*-test, $t=1.31<t_{0.05,385}=1.65$). The length at which fifty percent of the fish became mature was 47.41 cm TL (Fig. 4).

Males were found to be between 38 and 80 cm TL, females between 36 and 80 cm. The length range of

immature individuals was from 36 to 38 cm. The parameters of the total length to total weight relationship for males and females separately, and for the population as a whole, are given in Table 4. No significant difference in the allometric coefficient was found between males and females (*t*-test, $t=1.53<t_{0.05,771}=1.65$). Isometric growth was observed in both sexes and for the whole population (Table 4).

Of the otoliths examined, 670 (86.4%) were readable and used for the study of age and growth. The

APE value was only 3.4%. A false hyaline ring interrupting the normal growth pattern of the otolith was identified within the fourth annual opaque zone and in all subsequent opaque zones. Marginal zone analysis showed that one annulus was formed per year (Fig. 5). The percentage of otoliths with opaque edge was high in the months from April to September, and between June and August in particular.

Fish aged 3 to 11 years were present in the samples (Table 5). Growth parameters determined for males, females, and the entire population are shown in Table 6. No significant differences in the growth parameters were found between sexes (Hotelling´s T^2-test, T^2=5.29<$T_{0\ 0.05,3,\ 666}^2$=7.88).

The length converted catch curve is shown in Figure 6. The rates of total mortality, natural mortality, and fishing mortality were Z=0.49/year, M=0.35/year, and F=0.14/year, respectively. The size at first capture was 51.57 cm TL.

Discussion

Promethichthys prometheus is distributed along the slope to a depth of 800 m (Nakamura and Parin, 1993). In waters off the Canary Islands, greatest concentrations of this species are found between 400 and 700 m depth. Below this depth, the species is replaced by other trichiuroid fish present in the area, e.g. the black scabbardfish, *Aphanopus carbo* Lowe, 1839 (Uiblein et al., 1996).

The roudi escolar off the Canary archipelago is a gonochoristic species with no evidence of sexual dimorphism. The sex ratio is clearly unbalanced in favor of females. This fact could be explained by the differences between sexes in the spatial distribution. The lowest discrepancy between sexes observed during the reproductive season seems to confirm this conclusion. Because of the space partitioning between sexes and because females are fished more than males, this species could be classed as vulnerable to unrestrained fishing. Therefore, fishing for roudi escolar is an activity that has the potential to threaten its target population compared with fishing for more reliable and robust stocks (Csirke, 1988). This fishery will require a prudent exploitation strategy to reduce the potential risk of a collapse.

The roudi escolar has a definite reproductive period (extending from April to September, with a peak

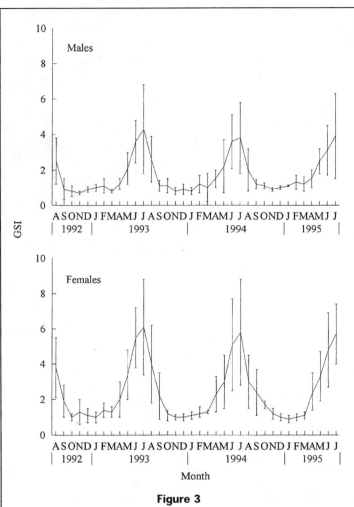

Figure 3

Monthly evolution of the gonadosomatic index (GSI) for males and females of *P. prometheus* caught off the Canary Islands (August 1992–July 1995).

Table 3

Number of males and females of *P. prometheus* caught off the Canary Islands by depth stratum and sex ratio tested by chi-square analysis. *=χ^2>$\chi_{t\ 1,0.05}^2$=3.84.

Depth (m)	Males	Females	Sex ratio	χ^2
201–300	4	26	1:6.50	16.13*
301–400	21	136	1:6.47	84.23*
401–500	35	187	1:5.34	104.07*
501–600	28	39	1:1.39	1.81
601–700	76	48	1.0.63	6.32*
701–800	85	41	1:0.48	17.12*
801–900	32	12	1.0.37	9.09*

in spawning activity in June–July) which agrees with information reported by Parin (1986) and Nakamura and Parin (1993). These authors pointed out that the

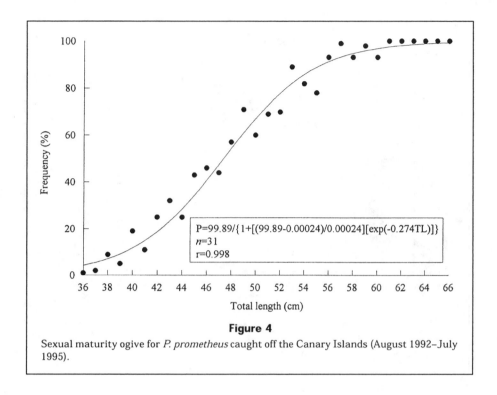

Figure 4

Sexual maturity ogive for *P. prometheus* caught off the Canary Islands (August 1992–July 1995).

spawning of this species occurs from August to September in the Atlantic off Madeira. Nishikawa (1987) found the highest concentration of gempylid larvae at the end of the summer months in the Pacific waters of Japan. In the Canary archipelago, the spawning of roudi escolar seems to be related to water temperatures, occurring when these reach greatest values. Possible significance of seasonal temperature variation to maturation and spawning in other bony fish species off the Canary Islands has been discussed by Lorenzo and Pajuelo (1996) and Pajuelo and Lorenzo (1995, 1996). Reproduction in roudi escolar does not involve horizontal migrations because during the spawning season the specimens are observed in the same areas where they are fished all year round. The lowest discrepancy in the number of males and females observed during the spawning period suggests an aggregation for breeding. Males and females were found aggregated mainly at depths of 450–650 m during this period.

Length at sexual maturity does not differ between males and females, corresponds approximately to 48 cm TL. In the age-length relationship, this size corresponds to 4-year-old specimens. The size at which fifty percent of the fish become mature is less than the length at first capture and the majority of the total catch is longer than this length, indicating a good exploitation pattern from the biological point of view. Furthermore, a low value of fishing mortality rate was obtained.

Table 4

Parameters of the length-weight relationship for males, females and all fish of *P. prometheus* caught off the Canary Islands and the possibility of isometry tested by Student's t-test. *= $t > t_{0.05, n > 250} = 1.65$.

	a	b	SE(b)	r^2	n	t-test
Males	0.004128	2.95321	0.03721	0.959	282	1.25
Females	0.004987	2.96214	0.03140	0.981	491	1.20
All fish	0.004521	2.98932	0.02341	0.992	776	0.45

The alternate pattern of opaque with translucent zones was easily distinguishable on the otoliths of the roudi escolar. These zones are deposited owing to alternating periods of rapid and slow growth (Williams and Bedford, 1974). The opaque zone is formed when the water temperature is higher, and food is abundant, and the translucent is formed when temperature is lower and the species spawns. This finding demonstrated the validity of using otoliths for estimating the age and growth of roudi escolar. The false hyaline zones observed within the fourth opaque ring and in subsequent opaque rings are probably spawning bands, because this species spawns in the summer months, when the opaque zone is formed in the otoliths. Morales-Nin (1987) pointed out that when the spawning does not take place during the period of hyaline zone formation, a false ring, known

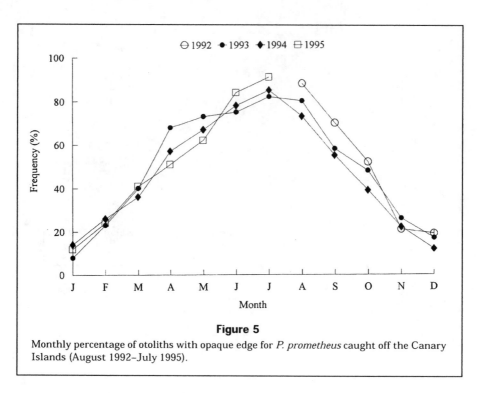

Figure 5

Monthly percentage of otoliths with opaque edge for *P. prometheus* caught off the Canary Islands (August 1992–July 1995).

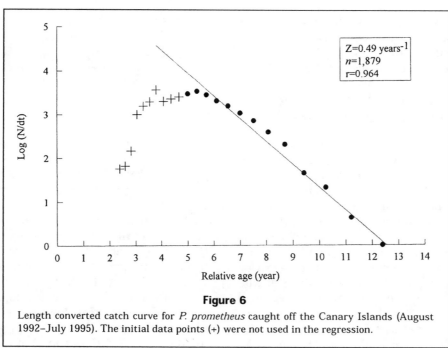

Figure 6

Length converted catch curve for *P. prometheus* caught off the Canary Islands (August 1992–July 1995). The initial data points (+) were not used in the regression.

as a spawning ring, may form within an opaque zone, dividing it into two.

The oldest age class observed was XI years, although this class, as well as age classes IX and X, were poorly represented in the landings. As a whole, growth of the roudi escolar is relatively fast and males and females grow at equal rates. The growth parameters obtained are reasonable because the theoretical maximal length value is greater than the size of the largest fish sampled and because the growth coefficient value indicates relatively rapid attainment of maximal size.

Table 5

Age-length key for all fish of *P. prometheus* caught off the Canary Islands.

Size (cm)	III	IV	V	VI	VII	VIII	IX	X	XI
36	2								
37	1								
38	2								
39	4								
40	6								
41	9								
42	9	1							
43	7	3							
44	4	4							
45	3	8							
46	3	7							
47		9							
48	2	13	1						
49		17	2						
50		14							
51		7	4						
52		5	6						
53		4	8						
54		3	10	1					
55			16	3					
56		2	22	3					
57		1	19	6					
58			12	7					
59			8	11	2				
60			4	17	2				
61			2	28	3				
62			3	21	6				
63			1	11	8				
64				8	14	2			
65				7	20	1			
66				3	23	4			
67				1	16	7			
68				2	10	10	1		
69					6	12	4		
70					3	17	4		
71					2	11	7	1	
72					1	6	9	2	
73						5	9	2	
74						1	7	3	1
75						2	5	6	1
76							2	4	2
77							1	3	2
78									1
79							1		
80									1
n	52	98	118	129	116	78	49	22	8
x	41.6	48.9	55.7	61.1	65.5	69.4	72.4	74.9	76.7
SD	2.4	2.8	2.6	2.6	2.4	2.2	2.0	1.8	1.7

Table 6

Parameters of the von Bertalanffy growth curve for males, females, and all fish of *P. prometheus* caught off the Canary Islands.

	L_∞ (cm)	k (per year)	t_0 (years)	r^2
Males	91.93	0.18	−1.66	0.964
Females	94.03	0.17	−1.58	0.978
All fish	93.61	0.18	−1.54	0.988

help with improving our English. This study was partially financed by the Directorate General XIV (Fisheries) of the European Commission.

Literature cited

Anderson, R. O., and S. J. Gutreuter.
  1983. Length, weight, and associated structural indices. *In* L. A. Nielsen and D. L. Johnson (eds.), Fisheries techniques, p. 283–300. Am. Fish. Soc., Bethesda, MD.

Beamish, R. J., and D. A. Fournier.
  1981. A method for comparing the precision of a set of age determination. Can. J. Fish. Aquat. Sci. 38:982–983.

Csirke, J.
  1988. Small schoaling pelagic fish stock. *In* J. A. Gulland (ed.), Fish population dynamics: the implications for management (second edition), p. 271–302. John Wiley & Sons, Chichester.

Gayanilo, F. C., Jr., M. Soriano, and D. Pauly.
  1988. A draft guide to the compleat ELEFAN. ICLARM Software 2, Contribution No. 435:1–65.

Lorenzo, J. M., and J. G. Pajuelo.
  1995. Biological parameters of the roudi escolar *Promethichthys prometheus* (Pisces: Gempylidae) off the Canary Islands. Fish. Res. 24:65–71.
  1996. Growth and reproductive biology of chub mackerel *Scomber japonicus* off the Canary Islands. S. Afr. J. Mar. Sci. 17:275–280.

Morales-Nin, B.
  1987. Métodos de determinación de la edad en los osteictios en base a estructuras de crecimiento. Inf. Téc. Inst. Inv. Pesq. 143:1–30.

Nakamura, I.
  1981. Gempylidae. *In* W. Fischer, G. Bianchi and W. B. Scott (eds.), FAO species identification sheets for fishery purposes. Vol II: Eastern central Atlantic; fishing areas 34, 47 (in part). FAO, Ottawa (unpaginated).

Nakamura, I., and N. V. Parin.
  1993. FAO species catalogue. Vol. 15: Snake mackerels and cutlassfishes of the world (families Gempylidae and Trichiuridae): an annotated and illustrated catalogue of the snake mackerels, snoeks, escolars, gemfishes, sackfishes, domine, oilfish, cutlassfishes, scabbardfishes, hairtails, and frostfishes known to date. FAO Fish. Synop. 125:1–136.

Nishikawa, Y.
  1987. Occurrence and distribution of gempylid larvae in the Pacific waters of Japan. Bull. Jpn Soc. Fish. Oceanogr. 51:1–8.

Acknowledgments

We are grateful to all persons who participated in sampling. We thank C. Tascón and N. Hernández for

Pajuelo J. G., and J. M. Lorenzo.

   1995.   Biological parameters reflecting the current state of the exploited pink dentex *Dentex gibbosus* (Pisces: Sparidae) population off the Canary Islands.   S. Afr. J. Mar. Sci. 16:311–319.

   1996.   Life history of the red porgy *Pagrus pagrus* (Teleostei: Sparidae) off the Canary Islands, central-east Atlantic. Fish. Res. 28:163–177.

Parin, N. V.

   1986.   Gempylidae.   *In* P. J. P. Whitehead, M. L. Bauchot, J. C. Hureau, J. Nielsen and E. Tortonese (eds.), Fishes of the north-eastern Atlantic and the Mediterranean, p. 967–973.   UNESCO, Paris.

Pauly, D.

   1980.   On the interrelationships between natural mortality, growth parameters and mean environmental temperature in 175 fish stocks.   J. Cons. Int. Explor. Mer 39: 175–192.

1983.   Length-converted catch curves: a powerful tool for fisheries research in the tropics (part I).   Fishbyte 1:9–13.

1984.   Length-converted catch curves: a powerful tool for fisheries research in the tropics (part II).   Fishbyte 2:17–19.

Ricker, W. E.

   1973.   Linear regression in fishery research.   J. Fish. Res. Board Can. 30:409–434.

Saila, S. B., C. W. Recksiek, and M. H. Prager.

   1988.   Basic fishery science programs.   Develop. Aquac. Fish. Sci. 18:1–230.

Uiblein, F., F. Bordes, and R. Castillo.

   1996.   Diversity, abundance and depth distribution of demersal deep-waters fishes off Lanzarote and Fuerteventura. Canary Islands.   J. Fish Biol. 49:75–90.

Williams, T., and B. Bedford.

   1974.   The use of otoliths for age determination.   *In* T. B. Bagenal (ed.), The ageing of fish, p. 114–123.   Unwin Brothers, Surrey.

"The Independence Gap and the Gender Gap," Barbara Norrander, *Public Opinion Quarterly*, Vol. 61, Issue 3, pp. 464-476.

Background

In the 1960s, the term "Generation Gap" was used to describe the difference in politics between the younger generation and the older generation. In this paper, the "gap" is the difference in the tendency of men and women to describe themselves as neither Democrat or Republican, but rather as Independent.

Questions
1. Table 1 contains data from the ANES surveys from 1952 to 1994. What do the numbers mean in the columns Democrat, Independent, and Republican?
2. Can you verify the calculations in the Z column? (There has been some rounding, so you may not get exactly the Z-values given.) What conclusion do the authors draw from Table 1? Why did they calculate Z-statistics for each year? What would have been the effect of combining years, say into 4-year periods of the presidential elections?
3. On p. 468, the author states that Table 2 demonstrates that the gap is located between the leaning-independent and weak-partisan categories. What statistical evidence is offered? What test could you do to confirm this result? The sample sizes in Table 2 are quite large. What is the effect of these large sample sizes? Is this a problem in Table 2? In the text, the author states that the pattern is statistically significant in (roughly) 3/4 of the surveys. Would you expect this many surveys to be significant, or more or fewer?
4. On p. 469, the author states that Table 3 shows that the independence gap is largest among the younger generation. Where is this result in Table 3? What would you say about this result?
5. The note at the bottom of Table 3 says that all patterns are statistically significant based on Pearson's chi-square. In Table 1, the author calculated a Z statistic. Does this mean that the analysis of Table 1 and Table 3 is different? What about the comment that "separated" is not statistically significant? What kind of analysis was done, do you think?
6. On p. 472, the author discusses how many groups men and women feel close to. A t-statistic of minus 3.10 is given, but no df. Does this mean that you cannot interpret the value? What about the practical significance of the difference? Do you think the assumptions of the t test would be valid? Do you know another approach to testing if men and women feel close to the same number of groups?

THE INDEPENDENCE GAP AND THE GENDER GAP

BARBARA NORRANDER

Abstract   An independence gap, in which women are on average 6 percentage points less likely than men to view themselves as political independents, has existed in the United States for at least 40 years. Women opt for weak partisanship, while men choose the leaning-independent category. Demographic differences between men and women do not explain this gap. Indirect evidence indicates it may be due to men placing a greater value on separateness and women placing a greater value on connections with others. When leaning independents are ignored, the partisan gender gap appears to be due mainly to women's greater attraction to the Democratic Party, but when these independents are folded in with the appropriate partisan group, the gender gap is seen to be due equally to mens' greater attraction for the Republicans.

Numerous scholarly studies and journalistic accounts focus on women's greater preference for the Democratic Party. These accounts pinpoint 1980 as the beginning of the gender gap. Yet, a longer standing difference between men's and women's partisan identities has been largely ignored. Since the 1950s, fewer women than men have called themselves independents. Bendyna and Lake (1993) note this difference but only as an explanation of how women could be both more Democratic and more Republican than men. Burnham (1970, p. 129) lists women as a group that is less likely to be independent than men but does not discuss it. Jennings and Niemi (1981, pp. 284–86) found young women in the early 1970s were less likely to be independents than were young men but concluded it was an anomaly specific to that time. But the independence gap is not an anomaly; it has persisted for 40 years and thus warrants more extensive investi-

BARBARA NORRANDER is associate professor in the Department of Political Science, University of Arizona. She wishes to thank Bill Lockwood and Richard Witmer for their assistance with the data analysis.

Public Opinion Quarterly Volume 61:464–476 © 1997 by the American Association for Public Opinion Research
All rights reserved. 0033-362X/97/6103-0004$02.50

gation. This article documents the independence gap from 1952 to 1994, demonstrates its persistence across a wide spectrum of subgroups, and investigates its consequences for the political behavior of men and women as well as its impact on the more familiar partisan gender gap.

How Large Is the Independence Gap?

The independence gap averaged 6 percentage points across the American National Election Studies (ANES) surveys, from 1952 to 1994, and was statistically significant in 16 of the 22 years. The size of the gap varied inconsistently during the 1950s and 1960s, grew stronger in the 1970s and 1980s, diminished in the early 1990s, but rebounded in 1994 (table 1).

The independence gap is also found in other surveys. The Jennings and Niemi (1981) socialization study observed it, as did Burnham's (1970) analysis of Gallup polls. The gap averages 5.6 percentage points in the General Social Surveys (GSS) from 1972 to 1994 and is statistically significant in 13 of the 20 years. Furthermore, in most years when the independence gap does not attain significance in the GSS (1972, 1973, 1974, 1980, 1984, 1989, and 1993) it is significant in the ANES surveys.[1] Thus, any comparison of the partisan preferences of men to women must account for gender variations in preferences for the independent label.

Is the Independence Gap a Measurement Phenomenon?

The ANES measure of partisanship is more complicated than a simple trichotomy of Democrat, independent, or Republican. Respondents are given the option of specifying another party, and some respondents are classified as apolitical because of responses to other questions that suggest a lack of interest in politics. If more women than men are classified as adherents of third parties or as apolitical, the percentage of women independents may be reduced. In fact, very few respondents claim identification with another party. Between 1988 and 1994, no more than 10 ANES respondents per survey claimed identification with a minor party. Thus, the number of third-party adherents cannot account for the independence gap.

A more likely possibility is that women are more often classified as apolitical than men. Women are more willing than men to reply ''don't know'' to survey questions (Rapoport 1981). Compared with women with

1. The 1974 GSS result is almost significant ($z = 1.95$).

Table 1. The Independence Gap, 1952–94

Year and Gender	Democrat	Independent	Republican	N	Independence Gap	Z
1952:						
M	48	26	26	801	5*	2.60
F	49	21	30	928		
1954:						
M	51	24	25	520	2	.75
F	48	22	31	568		
1956:						
M	46	28	26	767	6**	2.95
F	45	22	34	923		
1958:						
M	49	23	28	824	5**	2.39
F	53	18	30	913		
1960:						
M	43	29	28	858	10**	4.93
F	49	19	32	1006		
1962:						
M	50	24	27	567	4	1.52
F	47	20	32	670		
1964:						
M	51	26	24	689	5*	2.13
F	54	21	26	847		
1966:						
M	46	30	24	562	3	1.02
F	47	28	26	701		
1968:						
M	43	32	25	676	4	1.58
F	48	28	24	855		
1970:						
M	43	34	23	640	5*	1.98
F	45	29	26	850		
1972:						
M	37	39	23	1151	7**	3.96
F	44	32	24	1505		
1974:						
M	35	43	23	1040	9**	4.54
F	43	34	23	1393		
1976:						
M	37	43	21	1195	11**	5.73
F	42	32	26	1629		
1978:						
M	38	44	19	993	9**	4.43
F	42	34	23	1231		

Table 1 (*Continued*)

Year and Gender	Democrat	Independent	Republican	N	Independence Gap	Z
1980:						
M	38	40	22	679	8**	3.21
F	45	32	24	898		
1982:						
M	39	35	26	619	8**	3.01
F	50	27	23	764		
1984:						
M	34	39	28	968	7**	3.57
F	41	32	28	1230		
1986:						
M	37	38	25	926	7**	3.53
F	44	30	26	1194		
1988:						
M	30	42	29	856	9**	4.19
F	40	32	28	1143		
1990:						
M	36	36	27	881	2	.96
F	43	34	23	1054		
1992:						
M	32	40	28	1139	3	1.37
F	40	37	23	1309		
1994:						
M	30	38	33	824	5**	2.29
F	39	32	29	948		

Source.—ANES (1952–94).

Note.—Z = significance test for difference in proportions of men and women who claim independent identifications. Percentages may not match subtractions because of rounding.

* $p \leq .05$.

** $p \leq .01$.

little political interest, men with similarly low interest may be less willing to admit to not knowing and thus end up being classified as independents.[2]

2. The ANES coding for apoliticals has changed over the years. From 1952 to 1966, people who volunteered an answer to the first party identification question, which indicated "disinterest in politics," were immediately coded as apoliticals. From 1968 to 1980 and in 1984, those who responded to the first party identification question with "no preference" and to the independent-leaner probe with "neither" or "don't know" were classified as apolitical if they "otherwise also express a general disinterest in politics." In 1982 and since 1986, respondents are classified as apolitical if they answer "no preference," "neither," or "don't know" to the two party-identification questions and show little interest in politics

Indeed, women in the 1952 to 1962 ANES were slightly more likely than men to be listed as apolitical, by 3 percentage points. But after 1964, this difference averaged less than 1 percentage point.

Not all respondents who originally indicate no preference among the Republican, Democratic, and independent labels are classified as apolitical. Some become classified as pure independents or leaning independents (Miller and Wattenberg 1983). In the 1988 to 1994 ANES surveys, an average of 6 percent of men and 7 percent of women answered the initial Republican/Democrat/independent choice with "no preference." Thus, the independence gap, particularly after the early 1960s, is not due to more women than men being apolitical or having no party preference.

Where does the independence gap occur on the 7-point partisanship scale? There are two types of independents: those who lean toward the Democratic or Republican Party and pure independents. Similarly, two types of partisans exist: strong and weak. Folding the 7-point scale by these types of categories produces the 4-point partisan intensity scale: pure independents, leaning independents, weak partisans, and strong partisans. Whether the independence gap is concentrated among pure independents as opposed to among leaning independents could have different behavioral and theoretical consequences. *The American Voter* (Campbell et al. 1960) describes independents as less involved, less knowledgeable, and less interested in politics than partisans. Leaning independents, however, have often been shown to be as active as weak partisans. This intransitivity in the partisan intensity scale has lead some to theorize that leaning independents are closet partisans (Keith et al. 1992) and others to assert leaning independents are expressing their voting intentions (Shively 1980), while still others view leaning independents as true independents who happen to be more politically involved than other independents (Petrocik 1974).

Table 2 demonstrates that the gap is located between the leaning-independent and weak-partisan categories. Identical patterns emerge in both the 1952–94 ANES and 1972–94 GSS cumulative files. Women respondents are more likely to call themselves weak partisans, while more males opt for the leaning-independent category. This pattern is statistically significant in 16 of the 22 ANES surveys and 14 of the 20 GSS surveys.

Do Demographic Traits Explain the Independence Gap?

One explanation for why men and women answer partisanship questions differently would be demographic dissimilarities between the sexes.

---

in response to four specific questions on political interest (ANES 1952–94, coding for variable 301).

Table 2. Strength of Partisanship and Independence for Men and Women

	Men	Women	Difference
ANES cumulative file, 1952–94:			
Pure independents	13	13	0
Leaning independents	23	18	5
Weak partisans	34	39	−5
Strong partisans	30	30	0
No. of cases	18,418	23,156	
GSS cumulative file, 1972–94:			
Pure independents	13	12	1
Leaning independents	24	19	5
Weak partisans	38	43	−5
Strong partisans	26	26	0
No. of cases	13,893	17,880	

Women tend to live longer than men, and thus surveys contain more elderly women than men. These more numerous older women may simply express the more partisan preferences of past eras and provide women as a whole with a more partisan complexion. Another possibility would be educational differences. The lower educational levels of women may explain their propensity to avoid the independent label, since college-educated citizens have increasingly become attracted to nonpartisanship (Ladd and Hadley 1978). Conversely, the independence gap may disappear among racial or religious groups with strong identities with one party. The independence gap might also be more prevalent in the South. Southern women entered the electorate later than women in other parts of the country, with significant increases occurring in the 1960s (Cassel 1979). With less political tradition, Southern women may have been more willing to adopt the Republican label during the 1960s realignment than were their male counterparts who became independents instead. Finally, because a marital gap has existed since the 1970s with married individuals more likely to vote Republican (Weisberg 1987), a check will be made for any link between marriage and the independence gap.

Table 3 presents the independence gap for different demographic groups by use of the combined 1952–94 ANES data. Contrary to the generational explanation, the independence gap is largest among the younger generation. Similarly, the independence gap remains after controls for education and is larger among the better educated. A similar pattern exists for income levels (not shown), with an independence gap in the highest

Table 3. Size of the Independence Gap among Different Demographic Groups, 1952–94

	Pure Independent	Leaning Independent	Weak Partisan	Strong Partisan	No. of Cases
Age:					
Born after 1942	2	6	−7	0	12,940
Born 1942 and before	−1	5	−4	0	28,410
Education:					
Grade school	−5	3	−3	5	7,789
High school	1	5	−6	0	20,138
College	3	7	−4	−5	13,440
Race:					
White	0	5	−5	0	36,380
Black	0	5	−7	3	4,389
Hispanic	0	4	−9	5	1,142
Region:					
East	2	7	−7	−1	8,589
Midwest	0	5	−4	−2	11,380
South	−1	4	−6	2	13,577
West	0	5	−4	−1	6,898
Marital Status:					
Married	−1	4	−6	2	26,367
Single	4	5	−5	−3	4,794
Divorced	0	6	−7	1	2,832
Separated	0	6	−6	0	1,173
Widowed	−1	1	−2	2	4,361
Living together	−2	14	−10	−2	416
Religion:					
Protestant	−1	5	−4	1	27,624
Catholic	1	5	−5	0	9,400
Jewish	1	13	−11	−3	1,036
Other	2	2	−5	1	3,165

SOURCE.—ANES 1952–94.

NOTE.—The independence gap is measured such that a negative number indicates fewer men than women in a category. All patterns are statistically significant at .05 level based on Pearson's chi-square, except for separated (significance = .07) and widowed (significance = .24).

three categories and a weak partisanship gap for the two lowest income groups. The gap appears among all occupations (not shown) except farmers. The joint working situation of farm couples may lead to more closely matched political identities, or a general creed of independence among farmers may extend to both sexes' political identities.

The gap also exists among all racial and religious groups, though it is largest among Jews, Hispanics, and African Americans despite the Democratic preferences of these groups.[3] Region seems irrelevant, as is residence in cities versus suburbs versus rural areas (not shown). The gap is similar across categories of marital status except for the widowed (who show little gap) and those living together though not married (who show a larger gap). It is clear that the varying demographic traits of men and women do not explain the independence gap.

Do Party Attitudes Explain the Independence Gap?

Another possibility is that women are more partisan because they rate the parties more positively. This might be reflected in questions about whether one party comes closer to representing one's attitudes than does the other party. The ANES cumulative file has 11 proximity measures for respondents' perceptions of their own issue positions and those of the two parties. On eight issues, neither men nor women are more likely to see one party as closer to their own positions. On the three where gender differences occur (ideology, government services, and guaranteed job), men rather than women are more likely to view one party as closer to their own views.

Another measure of partisan attitudes is the series of open-ended questions concerning party likes and dislikes. Men have more responses to these questions in the ANES surveys than do women in both the positive and negative directions. Men average 1.1. positive and .9 negative comments about the Democratic Party and .9 and 1.0 negative remarks about the Republicans. Women average .9 positive and .6 negative remarks toward the Democratic Party and .7 and .8 negative toward the Republicans. (All gender differences are significant at the .01 level.) Thus, overall attitudes about the parties cannot explain the independence gap.

Gender Identities

Does the independence gap stem from a more general difference in the manner in which men and women view the political world? Gilligan

3. Hispanic identity was not measured prior to 1978. White and black categories shown do not attempt to separate out those of Hispanic backgrounds.

(1982) suggests that socialization leads men to value separateness and women to value connections with others. Men's preference for separateness might be expressed in political independence, while women's sense of belonging might lead them to partisanship.

Only a few crude tests of this idea can be undertaken, since political surveys usually do not measure such psychological orientations. When given the option of selecting a number of groups to which they feel close, women do not choose more groups than men. In the 1988 ANES survey, both women and men on average felt close to 3.5 groups. When categories for "women" and "feminists" are added to the list of groups, women respondents select 4.0 groups and men choose 3.8 ($t = -3.10$, $p < .01$).

While women may not indicate closeness to more groups, they rate a variety of groups in society higher than men. According to the 1952–94 ANES cumulative file, women rated 21 groups with an average feeling thermometer score of 63.7; men gave the cooler average rating of 60.8 ($t = -19.71$, $p < .01$).[4] Women rated 19 of the 21 groups higher than men, with the exceptions being big business and conservatives. These thermometer ratings provide some evidence that women feel more connected to a wider variety of groups than do men, which might help explain women's greater propensity to identify with a political party.

The Independence Gap and the Meaning of Partisanship

Does the difference in partisan intensity, reflected in the independence gap, affect men's and women's political behavior? In general, men are no more likely than women to split their tickets between the presidential and House of Representative contests. Only in one year (1980) of the 11 presidential elections covered in the 1952–94 ANES surveys did men split their tickets more than women. Similarly, there is little difference in the extent to which men and women split their tickets between the senatorial and presidential races. Again, only in one year (1992) did a gender difference occur, this time with women being more likely than men to split their tickets. Finally, the defection rate from party identification (including leaners as partisans) in presidential voting is typically the same for both sexes. The sole exceptions is 1988 when women were more likely than men to defect from their party in their presidential vote. In all three of these areas of voting behavior, men and women in the same category of partisan intensity scale have the same behaviors. The two exceptions occur for split-ticket voting between the U.S. House and the presidency. Here, among pure independents, men are more likely than women to split

4. These 21 groups exclude any references to women (i.e., women's liberation movement, women) and political parties.

their tickets, and among weak partisans women are more likely than men to split their ballots between the two parties.

Women are, however, less likely than men to report splitting their tickets among state and local races, on average by 6 percentage points. This pattern is statistically significant in 10 of 14 years. Women are also less likely than men to report that they have voted for different parties over time for president, by 7 percentage points. This pattern is statistically significant in 10 of 13 surveys. The greater partisan attachments among women may be causing them to be more partisan in their voting for the low-information state and local races and creating more loyalty over time to one party. However, women in all partisan categories, including pure independents, are more likely than men in the same category to report greater party loyalty across time in presidential contests. Women in both the leaning-independent and weak-partisan categories are more likely than men in those categories to report more straight-ticket voting for lower-level offices. Thus, gender differences in partisan intensity apparently do not underlie these small differences in electoral behavior.

Since partisan intensity theoretically leads to greater participation, the propensity of men to adopt an independent identification should decrease their political participation. Yet, men historically have participated at higher rates than women, for a variety of reasons. If men participate at slightly higher rates than women and are more likely to be leaning independents while women are more likely to be weak partisans, these gender patterns could be contributing to the intransitivities in the partisan identification scale (Petrocik 1974). This, however, is not the case. Controlling for sex does not eliminate intransitivities in turnout, the participation scale, interest in the current election, interest in public affairs, internal political efficacy, or media usage.

The Independence Gap and the Gender Gap

Does the independence gap play a role in the more traditional gender gap? Put differently, do the two different ways of coding leaning independents affect inferences about partisan differences between men and women?

Table 4 measures the partisan gender gap when independent leaners are combined with pure independents (left two columns) and when they are combined with partisans (right two columns). The figures on the left side suggest a greater attachment of women for the Democratic Party (negative values indicate more women prefer a party), a pattern that first emerges significantly in 1960 and stabilizes after 1972. However, the distinctiveness of gender patterns in partisan preferences is muted during the earlier period because, until the 1980s, women were also generally more likely than men to identify themselves as Republicans. Indeed, it is not

Table 4. Gender Gap Measured with Leaners Included and Excluded from Partisan Categories

	Leaners Classified as Independents		Leaners Classified as Partisans	
	Democrat	Republican	Democrat	Republican
1952	−2	−4	1	−3
1954	3	−5	5	−6*
1956	2	−8**	6*	−7**
1958	−3	−1	1	−1
1960	−6**	−4	1	−2
1962	2	−6*	4	−6*
1964	−3	−2	0	0
1966	−1	−2	−1	1
1968	−5*	2	−4	4
1970	−3	−2	0	0
1972	−6**	−1	−5**	3
1974	−9**	−1	−5*	3
1976	−5**	−5**	−2	−1
1978	−4*	−5**	−3	−1
1980	−7**	−1	−6*	3
1982	−10**	3	−9**	6*
1984	−8**	0	−5*	4
1986	−6**	−1	−5*	5*
1988	−10**	1	−8**	7**
1990	−6**	4*	−4	6**
1992	−8**	5**	−10**	8**
1994	−9**	4	−10**	9**
1952–62	−1	−5**	3**	−4**
1964–78	−5**	−2**	−3**	1
1980–94	−8**	2**	−7**	6**

SOURCE.—ANES 1952–94.

NOTE.—Values are percentage points differences between men and women in the category. Positive number indicates more male respondents.

* $p \leq .05$.

** $p \leq .01$.

until 1990 that the pattern reverses and men are significantly more apt to identify themselves as Republicans.

When leaners are treated as partisans as opposed to independents (right side of table 4), the patterns are somewhat different. On the basis of this fuller information, it is men who are more apt to identify with the Democrats in the 1950s and early 1960s, not women. Moreover, the greater propensity of men to identify as Republicans now becomes clearly established in the very early 1980s, not 1990. Thus, although more traditional treatments of the gender gap (which ignore leaning independents) mainly highlight women's greater concentration among Democrats, including the leaners reveals an equally great concentration of men among Republicans.

Conclusions

An independence gap, in which women are on average 6 percentage points less likely than men to view themselves as political independents, has existed in the United States for at least 40 years. Women opt for weak partisanship, while men choose the leaning-independent category. Demographic differences between men and women do not explain this gap. Indirect evidence indicates it may be due to men placing a greater value on separateness and women placing a greater value on connections with others. When leaning independents are ignored, the partisan gender gap appears to be due mainly to women's greater attraction to the Democratic Party, but taking them into account shows the gap is due equally to mens' greater attraction for the Republicans. Thus, the independence gap can affect our understanding of an important aspect of American politics.

References

Bendyna, Mary E., and Celinda C. Lake. 1993. "Gender and Voting in the 1992 Presidential Election." In *The Year of the Woman: Myths and Realities,* ed. Elizabeth Adell Cook, Sue Thomas, and Clyde Wilcox, pp. 237–54. Boulder, CO: Westview.

Burnham, Walter Dean. 1970. *Critical Elections and the Mainsprings of American Politics.* New York: Norton.

Campbell, Angus, Philip E. Converse, Warren E. Miller, and Donald E. Stokes. 1960. *The American Voter.* New York: Wiley.

Cassel, Carol A. 1979. "Change in Electoral Participation in the South." *Journal of Politics* 41:907–17.

Gilligan, Carol. 1982. *In a Different Voice.* Cambridge, MA: Harvard University Press.

Jennings, M. Kent, and Richard G. Niemi. 1981. *Generations and Politics.* Princeton, NJ: Princeton University Press.

Keith, Bruce E., David B. Magleby, Candice J. Nelson, Elizabeth Orr, Mark C. Westlye, and Raymond E. Wolfinger. 1992. *The Myth of the Independent Voter.* Berkeley and Los Angeles: University of California Press.

Ladd, Everett Carll, Jr., and Charles D. Hadley. 1978. *Transformations of the American Party System.* 2d ed. New York: Norton.

Miller, Arthur H., and Martin P. Wattenberg. 1983. "Measuring Party Identification: Independent or No Partisan Preference." *American Journal of Political Science* 27: 106–21.

Petrocik, John R. 1974. "An Analysis of Intransitivities in the Index of Party Identification." *Political Methodology* 1:31–48.

Rapoport, Ronald B. 1981. "The Sex Gap in Political Persuading: Where the 'Structuring Principle' Works." *American Journal of Political Science* 25:32–48.

Shively, W. Phillips. 1980. "The Nature of Party Identification: A Review of Recent Developments." In *The Electorate Reconsidered,* ed. John C. Pierce and John L. Sullivan, pp. 219–36. Beverly Hills, CA: Sage.

Weisberg, Herbert F. 1987. "The Demographics of a New Voting Gap: Marital Differences in American Voting." *Public Opinion Quarterly* 51:335–43.

"Moxibustion for Correction of Breech Presentation – A Randomized Controlled Trial," Francesco Cardini, MD, Huang Weixin, MD, *JAMA*, Vol. 280, No. 8.

Background
Breech presentation is where the unborn baby is poised to enter the birth canal feet first. This is a complication that may be extremely serious for mother and baby, alike. "Moxibustion" is a traditional Chinese remedy. It involves burning a roll of herbs (sometimes described as rather like a cigar) near the mother's big toe. You might want to use the Web to find out more about moxibustion. (I have not been able to determine if it is performed with the mother standing or lying down.)

This article uses some terms that may not be familiar.
1. Under Patients, they use the term "primigravidas," which means first pregnancy. The patients were in their 33rd week. "Normal" pregnancy is 40 weeks. Birth before 35 weeks is considered "premature."
2. Under Interventions, "external cephalic version" is mentioned. This is where the doctor tries to move the baby by pushing on the mother's stomach to move the head so the baby is head-down.
3. Under Main Outcome Measures, they mention "cephalic presentations." This means the baby is head first, which is the safer way to give birth.

Questions
1. Under Design, the authors list "randomized, controlled, open clinical trial." How does "randomized" enter into this study? Why would it be important? Does this tell you if this study is an experiment or an observational study? In this study, what does the term "controlled" mean? What role does this play in the study?
2. Under Main Outcome Measures, they mention fetal movements counted 1 hour each day for 1 week. It does not say if the same hour is used each day. How would this be important? Would it be OK if the mother picked an hour each day on her own? Suppose the researchers gave the mother a randomly chosen hour each day to use. What would be the advantages and disadvantages of this?
3. Under results, they give both a p-value and a confidence interval for the difference in the mean number of fetal movements. Does the confidence interval seem reasonable for such a small p-value? Can you reconstruct either of these calculations? What method do you think they used? What assumptions are needed for this method? Is it reasonable to think these assumptions are satisfied? If you were doing the study, how would you go about determining if the assumptions are reasonable? Do you know any alternative methods if you are not comfortable with the necessary assumptions?
4. Comparing the probability of cephalic presentation at the 35th week, the authors use relative risk [RR]. If you are not familiar with this term, try to figure out what it means. (Hint: in this study, "risk" is not a bad thing.) They give a confidence interval for RR. What would the null hypothesis be in terms of RR? In light of this, does their confidence interval match up with their p-value? How would you explain the RR to a news reporter?
5. Calculate the exact p-value for the comparison of cephalic presentations (head first) at the 35th week and also check the given p-value for presentation at birth. Did the

authors use one-sided or two-sided p-values? Do you agree with their choice? If not, how would you determine your p-value in terms of their p-value?

6. Do the authors' conclusions match the number of sides they used in their objectives?

JAMA & ARCHIVES

The Journal of the American Medical Association —— To Promote the Science and Art of Medicine and the Betterment of the Public Health

Select Journal or Resource | GO

SEARCH THIS JOURNAL:

GO TO ADVANCED SEARCH >

HOME   CURRENT ISSUE   PAST ISSUES   COLLECTIONS   CONTACT US   HELP

Vol. 280 No. 18, November 11, 1998
Original Contribution

TABLE OF CONTENTS >

Moxibustion for Correction of Breech Presentation

A Randomized Controlled Trial

Francesco Cardini, MD; Huang Weixin, MD

JAMA. 1998;280:1580-1584.

Context.— Traditional Chinese medicine uses moxibustion (burning herbs to stimulate acupuncture points) of acupoint BL 67 (Zhiyin, located beside the outer corner of the fifth toenail), to promote version of fetuses in breech presentation. Its effect may be through increasing fetal activity. However, no randomized controlled trial has evaluated the efficacy of this therapy.

Objective.— To evaluate the efficacy and safety of moxibustion on acupoint BL 67 to increase fetal activity and correct breech presentation.

Design.— Randomized, controlled, open clinical trial.

Setting.— Outpatient departments of the Women's Hospital of Jiangxi Province, Nanchang, and Jiujiang Women's and Children's Hospital in the People's Republic of China.

Patients.— Primigravidas in the 33rd week of gestation with normal pregnancy and an ultrasound diagnosis of breech presentation.

Interventions.— The 130 subjects randomized to the intervention group received stimulation of acupoint BL 67 by *moxa* (Japanese term for *Artemisia vulgaris*) rolls for 7 days, with treatment for an additional 7 days if the fetus persisted in the breech presentation. The 130 subjects randomized to the control group received routine care but no interventions for breech presentation. Subjects with persistent breech presentation after 2 weeks of treatment could undergo external cephalic version anytime between 35 weeks' gestation and delivery.

Main Outcome Measures.— Fetal movements counted by the mother during 1 hour each day for 1 week; number of cephalic presentations during the 35th week and at delivery.

Results.— The intervention group experienced a mean of 48.45 fetal movements vs 35.35 in the control group ($P<.001$; 95% confidence interval [CI] for difference, 10.56-15.60). During the 35th week of gestation, 98 (75.4%) of 130 fetuses in the intervention group were cephalic vs 62 (47.7%) of 130 fetuses in the control group ($P<.001$; relative risk [RR], 1.58; 95% CI, 1.29-1.94). Despite the fact that 24 subjects in the control group and 1 subject in the intervention group underwent external cephalic version, 98 (75.4%) of the 130 fetuses in the intervention group were cephalic at birth vs 81 (62.3%) of the 130 fetuses in the control group ($P=.02$; RR, 1.21; 95% CI, 1.02-1.43).

Conclusion.— Among primigravidas with breech presentation during the 33rd week of gestation, moxibustion for 1 to

Featured Link
· **E-mail Alerts**

Article Options
· Full text
· PDF
· Send to a Friend
· Related articles in this issue
· Similar articles in this journal

Literature Track
· Add to File Drawer
· Download to Citation Manager
· PubMed citation
· Articles in PubMed by
  ·Cardini F
  ·Weixin H
· ISI Web of Science (29)
· Contact me when this article is cited

Topic Collections
· Pregnancy and Breast Feeding
· Complementary and Alternative Medicine
· Collection E-mail Alerts

2 weeks increased fetal activity during the treatment period and cephalic presentation after the treatment period and at delivery.

From the Jiangxi Women's Hospital, Nanchang, People's Republic of China (Dr Weixin). Dr Cardini is in private practice in Verona, Italy.

RELATED ARTICLES IN *JAMA*

This Week in *JAMA*
JAMA. 1998;280:1551.
FULL TEXT

"Undertreatment of Osteoporosis in Men with Hip Fracture," Gary Kiebzak, PhD, Garth A. Beinart, MD, Karen Perser, BA, Catherine G. Ambrose, PhD, Sherwin J. Siff, MD, Michael H. Heggeness, MD, PhD, *Archives of Internal Medicine* Vol. 162, No. 19.

Background
Osteoporosis is a condition where the bones lose their strength. It is often associated with old age. It can lead to bone fractures, which not only can affect mobility, but can actually be life threatening. Traditionally, osteoporosis is thought of as affecting women. This study considers if the treatment that men receive is equal to that received by women, and, if not, what are the results of this treatment difference?

Questions
1. Is this an experiment or an observational study? What would be the issues in constructing this as an experiment?
2. What is the basis for thinking that the men and the women in this study are comparable? That is, why does it make sense to compare the outcomes for the two groups?
3. The article states that at hospital discharge, the p-value for testing if men and women had the same probability of receiving any treatment for osteoporosis is < 0.001. What is the actual value of the p-value? Is merely saying "$P < .001$" sufficient? Construct a confidence interval to see how much less the probability of treatment for men is than the probability of treatment for women.
4. The p-value for testing equal mortality is given as 0.003. Is this a one-sided or a two-sided p-value? Which one is more appropriate? If you think the number of sides is wrong, how would you convert the given p-value to the one you believe is better?
5. At follow-up, the p-value for testing if men and women had different probabilities of receiving osteoporosis treatment is given as "$< .001$." How would you explain this to a news reporter? In doing a follow-up, it is not always possible to locate all the subjects. Does this present a problem in the given study? What lurking variables might there be?
6. At the follow-up, the fraction of men and women who had bone mineral density measurement was compared. Why is this part of this study? No p-value was given. If possible, calculate it and state your conclusion.

JAMA & ARCHIVES

Select Journal or Resource   GO

ARCHIVES OF
INTERNAL MEDICINE

SEARCH THIS JOURNAL:

GO TO ADVANCED SEARCH >

HOME | CURRENT ISSUE | PAST ISSUES | COLLECTIONS | CONTACT US | HELP

Vol. 162 No. 19, October 28, 2002

TABLE OF CONTENTS >

Original Investigation

Undertreatment of Osteoporosis in Men With Hip Fracture

Gary M. Kiebzak, PhD; Garth A. Beinart, MD; Karen Perser, BA; Catherine G. Ambrose, PhD; Sherwin J. Siff, MD; Michael H. Heggeness, MD, PhD

Arch Intern Med. 2002;162:2217-2222.

Background Women are not aggressively treated for osteoporosis after hip fracture; the treatment status of men with hip fracture has not been extensively studied.

Objective To evaluate the outcome and treatment status of men with hip fracture.

Methods Data from medical records were obtained for 363 patients (110 men and 253 women) aged 50 years and older with atraumatic (low-energy) hip fracture who were admitted to St Luke's Episcopal Hospital between January 1, 1996, and December 31, 2000. Surveys were mailed to surviving patients. Main outcome variables were osteoporosis treatments (antiresorptive or calcium and vitamin D) at hospital discharge, current osteoporosis treatments at 1- to 5-year follow-up, bone mineral density testing, mortality, current disability, and living arrangements (home or institution).

Results The mean age for men was 80 years vs 81 years for women. Most fractures (89% for men and 93% for women) resulted from falls from a standing height. At hospital discharge, 4.5% of men (n = 5) had treatment of any kind for osteoporosis, compared with 27% of women (n = 69) ($P<.001$). The 12-month mortality was 32% in men, compared with 17% in women ($P = .003$). Surveys were usable from 168 (87%) of 194 survivors. At 1- to 5-year follow-up, 27% (12/44) of men were taking treatment of any kind for osteoporosis, compared with 71% (88/124) of women ($P<.001$). Of those treated, 67% (8/12) of men and 32% (28/88) of women were taking calcium and vitamin D only. At 1- to 5-year follow-up, 11% of men had a bone mineral density measurement, compared with 27% of women. After hospital discharge, the number of men and women who required wheelchairs, walkers, and canes and who lived in institutions increased significantly.

Conclusions The burden of hip fracture is illustrated by the high incidence of postfracture disability and the high mortality rate in both men and women. Nevertheless, few men receive antiresorptive treatment.

From the Center for Orthopaedic Research and Education (Drs Kiebzak, Siff, and Heggeness) and Department of Orthopaedic Surgery (Dr Siff), St Luke's Episcopal Hospital; Department of Orthopaedic Surgery, Baylor College of Medicine (Drs Kiebzak, Siff, and Heggeness), Houston, Tex; The University of Texas Houston Medical School (Ms Perser); and Department of Orthopaedics, The University of Texas Houston Health Science Center (Dr Ambrose). During the preparation of this article, Dr Beinart was in medical school at Baylor College of Medicine. He is now affiliated with the Department of Internal Medicine, University of California, San Francisco.

Featured Link
· E-mail Alerts

Article Options
· Full text
· PDF
· Send to a Friend
· Related articles in this issue
· Similar articles in this journal

Literature Track
· Add to File Drawer
· Download to Citation Manager
· PubMed citation
· Articles in PubMed by
  · Kiebzak GM
  · Heggeness MH
· Contact me when this article is cited

Topic Collections
· Quality of Life
· Osteoporosis
· Men's Health
· Aging/ Geriatrics
· Collection E-mail Alerts

RELATED ARTICLES IN *ARCHIVES OF INTERNAL MEDICINE*

In This Issue of *Archives of Internal Medicine*
Arch Intern Med. 2002;162:2160.
FULL TEXT